# The Glider Pilot Badge 1940 - 1944

by

Stijn David

First Edition 2008



English text by Dietrich Maerz
Edited and revised by George Stimson

Cover layout by Robert Zatorski



**B&D Publishing, LLC**
**PO Box 652**
**Richmond, MI 48062**
**www.ironcross1939.com**

**ISBN: 978-0-9797969-3-7**

FOR MY WIFE SILVIA

Unknown private wearing the *Luftwaffe* Glider Pilot Badge (tropical cloth version) on his tropical uniform.

# FOREWORD

*Die Literatur zur Geschichte der deutschen Luftwaffe in den Jahren 1933 bis 1945 ist umfangreich und vielfältig. Trotzdem gibt es noch spezielle Themen, die bisher wenig oder gar keine Beachtung gefunden haben. Eines davon wird in dem vorliegenden Buch ausgiebig behandelt, wobei der Autor mit akribischem Eifer, profunder Sach- und Detailkenntnis und einem enormen Zeitaufwand, ein Kompendium erstellt hat, dem ich die gebührende Beachtung und Verbreitung wünsche.*

"The available literature about the German Luftwaffe between the years of 1933 - 1945 is comprehensive and manifold. However, there are still areas which have received only minor attention or none at all. One on those areas is addressed in depth with this book, in which the author has compiled a compendium with painstaking diligence, profound experience, detailed knowledge and with an enormous expenditure of time. I wish this book the due attention and propagation it deserves."

Helmut Schultz
Former glider pilot (DFS 230/Go 242/Me 321)

---

"*Vergessen heißt Erfahrungen aus dem Fenster werfen.*"
*zitiert nach Arthur Schopenhauer (1788 - 1860)*

*Dem Freund Stijn David für seine großartige Zusammestellung aus der Geschichte der deutschen Lastensegler. Hochachtung und Herzlichen Dank !*

"To forget is like throwing experience out of the window"
quoted from Arthur Schopenhauer (1788 - 1860)

To the friend Stijn David for his outstanding compilation of the history of the German glider troops! Our respect and cordial thanks!

Siegfried Baier & Anton Röhrle
(Glider troop veterans)

# TABLE OF CONTENTS

# ACKNOWLEDGEMENT

First and foremost I would like to thank my wife Silvia. Without her support and patience during the long hours of research this book would not have been possible.

I would also like to thank the following persons who supported me a great deal during my studies:

Mitglieder der Luftlande-Flieger-Kameradschaft
Frau Jesch
Ailsby, Christopher
Baier, Siegfried "Sigi"
Behnke, Hans
Beltram, Jody
Boschmann, Hans
Brücker, Bastian
Burgdorf, Dirk
Burtonhaw , Ian
Bühler, Pius
Couwenberg, Bruno
Döbel, Bernd
Eck, Robert (+)
Falkensteiner, Erwin
Garlasco, Marc
Gerstmann, Roswitta
Green, Jeremy
Grimm, Wolfgang
Hamma, Doris
Hatwell, Paul
Horn, Bruno
Jacques
Jauch, Willi
Jewison, Ian
Keating, Prosper
Köhler, Walter
Kortum, Karl - Heinz (+)
Lauser, Günter
Matthiesen, Walter
Oberle, Günther
Otto, Wolfgang
Petz, Bill
Powilleit, Heinz (+)
Reichert, Adam
Röhrle Anton, "Toni"
Schlaug, Georg
Schneider, Karl
Schneider, Karl - Heinz "Hati"
Schultz, Helmuth
Schurr, Hans "Bums" (+)
Söhnel, Heinz
Solbrig, Emil
Spiller, Walter
Ullrich, Erich
Verstraeten, Peter
Tannahill, Ian
Carsten Baldes / Ulm
Claus Phillip / Buchholz
Helmut Weitze / Hamburg
Carsten Zeige / Hamburg
Bundesarchiv-Militärarchiv Freiburg
SKF vzw.
Bund Deutscher Ordenssammler e.V.

all the anonymous collectors who helped

# INTRODUCTION

The Glider Pilot Badge is among the rarest of all the qualification badges of the *Luftwaffe* of the Third Reich.

Still today information about this badge is scarce. The infrequent and often contradictory information in the available publications, and the very low number of original examples of the award, do not make it very easy for the collector to distinguish between contemporary originals and copies (respectively, fakes).

This book tries to provide the interested collector with an overview of the history of this rare badge and should also serve as a collector's guide line.

The author wants to dedicate this book to the former glider pilots. They went to war 65 years ago for what they thought was a good cause, and a great many of enthusiastic pilots lost their lives while doing their duty. These men, and all the others which died in the Second World War, should not be forgotten.

The book includes only badges with absolute solid provenance. Variants, which have surfaced inside the collector community and for which no proof of originality has been provided or found so far, are consequently not included.

Stijn David

Tielt, Belgium
January 2008

# INTRODUCTION TO THE HISTORY OF THE GERMAN GLIDER TROOPS

To understand the history of the Glider Pilot Badge one also has to have some know-ledge about the structure of the *Luftwaffe* (German Air Force). After the introduction of the general conscription and the unmasking of the *Luftwaffe* in the spring of 1935 it was not too-far fetched to develop a transport glider which could serve as an observatory platform - just like the giant glider "Obs" of 1930 - as well as a transport of soldiers and weapons. In the summer of 1936 engineer Jacobs, head of the design department of the *Deutsche Forschungsanstalt für Segelflugzeuge* (DFS - German Glider Research Institute) in Darmstadt was charged to develop such an airplane.

As early as September 1936 a mock-up of the para glider (*Lastensegler LS*) model DFS 230 was shown to Ernst Udet and his staff. The first trial glider was supposed to be ready to test by March of 1937, but the research institute succeeded to conduct the first successful flight with the LS DFS 230 as early as January 1937.

The actual hour of birth of the German air-borne glider troops was the 11. January 1937. It was on that day that two trained paratroopers, *Leutnant* Kiess and *Unteroffizier* Flucke were detached to the DFS to learn para gliding flight during a three month course, and also to develop and test training methods for later instruction courses.

In accordance with a decree from the *Reichsluftfahrtsministeriums* (Air Force Ministry) a para glider-command was created in Darmstadt on 1. April, 1937, and only a short time later the first training course with 80 student flyers started. After the successful realization of this first seminar, a second course was started immediately (Picture below).

On the 16th of November 1937 a maneuver was held in Stendal/Saxony-Anhalt to compare the performances between paratroopers jumping with a parachute and paratroopers landing with a para glider DFS 230. This exercise was a complete success for the para gliders, and accordingly the para glider concept had the final break through. The *Reichsluftfahrtsministerium* (RLM, Air Force Ministry) ordered on the 9. March 1938 to establish (effective on 1. April 1938) the "*Ausbildungskommando für Lastensegelflug*" (Training Command for Para Gliders) in Fürstenwalde/Brandenburg, about 35 miles east of Berlin.

*Generalmajor* Student assumed the command of the paratroopers on 6th July 1938 and was also charged to form the *Fliegerdivision 7* (Air Force Division). The para gliders were further developed under this officer and a first military mission was contemplated during the so called "*Sudentenkrise*" (crisis of the Sudeten land) at the end of 1938. In particular the gliders were supposed to be used against the border fortifications between Germany and Czechoslovakia. Since the crisis was resolved without using military force, this mission never took place.

But twenty months later, on 10. May 1940, it was the first time in military history that para gliders were used in anger. The Belgian border fort *Eben Emael* was eliminated with a surprise attack conducted with para gliders. The detailed course of action can be assumed as well known to the reader. Para gliders landed on top of the fort and only a few paratroopers (about 70) succeeded in holding about 1200 soldiers of the fort at bay and also destroyed the main weapon cupolas of the heavily armed (and also assumed to be impregnable) border fortification.

The picture shows a DFS 230-A1 in rigid-tow behind a Junker Ju-52/3m

This outstanding success of the paratroopers was of course used by the German propaganda to the maximum effect. However, the crucial and decisive use of the para gliders was kept secret and was therefore not made public.

The picture (above) shows Hitler chatting with *Oberleutnant* Meissner, *Hauptmann* Witzig and *Mayor* Koch after awarding the *Ritterkreuz* to these paratroopers of the Eben Emael operation.

The leadership of the *Luftwaffe* immediately decided after this extremely successful maiden deployment of the para gliders during the *Fall Gelb* (Case Yellow - the attack on Belgium, the Netherlands and France) on the expansion of this new weapon. On 27. July 1940 the order was given to form the *Luftlandegeschwader 1* (LLG. 1 or Airborne Troop Squadron 1). The heart of the LLG 1 was the 17./K.G.z.b.V. 5

The view out of the cockpit of the DFS 230 glider under tow by a Ju 52.

# A Brief History Of The Glider Missions In World War 2

After the successful conclusion of the *Westfeldzug* (western campaign) there was a small break in active warfare on the European continent. However, for the now expanding glider forces this break did not mean the end of this new (and successful) weapon. On the contrary, it is around this time that quite a few young (civil) qualified glider flyers where drafted into the *Luftwaffe* to start their training to become a *Luftwaffe* glider pilot.

Beginning in 1941 the German forces focused their attention towards the south of Europe (*Balkan*) and then ultimately towards the east (Russia).

From that point on the glider force had to develop its skills into two different directions. After the initial phase of the invasion of Russia (*Operation Barbarossa*) the pure assault missions in which the glider force was involved lost quite a bit of terrain, and they had to take on a supportive role in flying emergency missions *(Notlösungen*), and were therefore used more and more as transport troops.

Going into detail into all the different missions the glider force was involved is not the goal of this book. However, for the readers interest a few pure "assault" and "transport" missions will be highlighted. The lesser known missions were chosen, as these are often equally interesting, and because the other "big" missions, such as the invasion of Crete (*Operation Merkur*) are already generally known and well covered in the literature.

Assault Missions:

- Ösel (Russia, 1941)
- *Kommandounternehmen* Tunesia (1942)
- Vassieux (France, 1944)

Transport Missions:

- Suschinitschi (Russia, 1941/1942)
- Welikije Luki (Russia, 1942/1943)
- Berlin (Germany, 1945)

The German glider troops where involved in quite a few more missions than the ones listed above, but these give an idea about the use of these forces during WW2. The author has also chosen not only a few successful missions but also some unsuccessful missions to show that not all actions ended in a victory.

# The Island of Ösel (Eastern Front 1941)

After the German invasion in Russia the German army group "North" very easily conquered the Baltics, but the German high command saw also a need to occupy the small Baltic islands which where still held by Russian troops. Therefore a combination of several naval, aerial and ground forces were assembled to conduct the task.

The island Ösel had a well fortified coastal area which was to be knocked out through an aerial landing. With this in mind, part of the 6./LLG.1 flew from Halberstadt/Brunswick to Seerappen (East Prussia).

A first attack wave of five DFS230 glider planes, each loaded with 9 Pioneers, had to attack the coastal defenses by surprise from the air. At the same time a sea borne force would land and bring additional forces to the island to occupy it successfully.

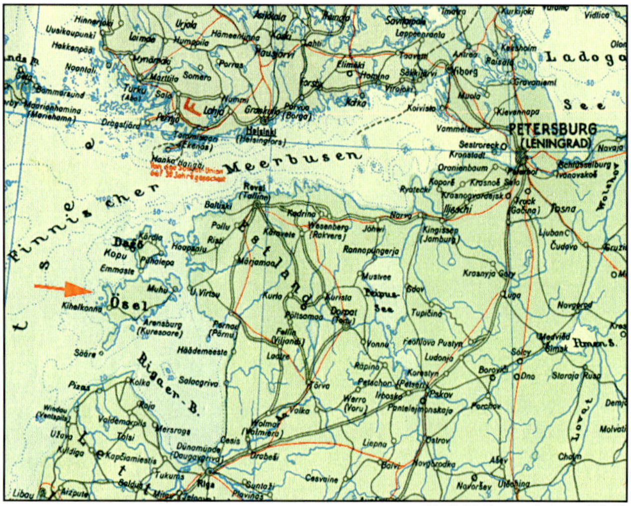

IRO Kriegs-Atlas, 1940

In the early morning of 14. September 1941 the gliders where loaded and towed towards their destination. After a 45 minute flight they reached their goal, and during the first attempt to release the gliders the aerial force received heavy Flak fire. A second attempt was necessary, and this time the gliders were safely released from the towing planes. The initial landing was a success and all five gliders managed to land safely without any damage. The only problem was that the five gliders did not land at the planned spot!

After the German soldiers (about 44) successfully disembarked they started the ground assault mission in advancing toward the Russian positions. In the beginning, the assault force thought that the enemy was weak and that they would be able to achieve their task without help. But that idea changed quickly when the enemy fire became stronger with dawn, and soon the Germans received fire from the left, right, and front. To make it worse they simply were not able to identify the well camouflaged bunkers. However, the assault force was still able to advance.

But the Russian defenders brought in fresh troops and managed to encircle the German force. This resulted in a situation where the attackers now had to defend themselves and due to the increasing opponents they suffered more losses resulting in a less effective assault force.

After some time the attackers were worn down to an effective fighting force of 30 soldiers and they had to defend themselves against serious attacks. The Germans managed to hold on until the evening when they were forced to draw back towards the coast.

Three Junkers Ju 52 transport planes dropped nine small rubber boats on the coast to support the withdrawal. Four of the boats fell into Russian hands so only five could be recovered. In the evening dusk the heavily wounded soldiers were placed into the boats and they set out towards the sea.

This nightly trip was also not without dangers. Almost all of the rubber boats were somewhat damaged, so in order to keep them afloat it was needed to pump out the incoming water. It was also decided in several boats that only two soldiers could stay inside and that the others had to swim along side. Soon all the boats were scattered and each boat was on its own. In the morning the soldiers from four boats were rescued by German vessels. However, the fifth boat with its nine soldiers could not be found.

Later it became clear what had happened. This boat was not seaworthy enough and had to return to land. As soon as the nine soldiers reached the shore of the island Ösel they were captured and became Soviet POWs. In the morning of 15. September 1941 these prisoners were shot by the Russian soldiers. (Only two of these nine German prisoners survived this war crime).

The Mission against the island of Ösel is an example of an unsuccessful glider assault. Later it was discovered that the 44 German soldiers had to fight against an overwhelming occupying force of 800-1000. Out of the initial 44 attackers, 11 German soldiers were killed during this operation and nine were wounded.

Gotha Go-242 Transport Glider

## *Kommandounternehmen* Tunesia (December 1942)

In mid-December 1942 the XI *Fliegerkorps* received the request from the *Oberbefehlshaber Süd* (Supreme Commander South) to detach an officer who was capable to plan and conduct commando missions against the supply routes used by the allied troops in Tunesia.

The mission's goal was to destroy bridges in the French and English occupied areas of Tunesia. To achieve this mission it was thought to either use German paratroopers or to land troops with glider planes. The overall commander of this mission was *Hauptmann* Schacht. To conduct the mission he received the following units:

- *1./Korps-Fallschirmpionierbataillon*
- *Abteilung Von Koenen (4.Rgt. "Brandenburg")*

In addition the *Luftwaffe* allocated six Junkers Ju 52 transport planes which had to transfer 12 DFS 230s from Lecce (Italy) towards Bizerta (Tunesia). Also a few He 111 were made available to join the mission as towing planes. The picture below shows a DFS 230 in Italy 1942.

A DFS 230 in Italy, 1942

The glider planes were towed on 26. December 1941 from Sicily to Bizerta. On the evening of the same day the first mission briefing took place with the commanding officers of the *Abteilung Von Koenen.* The final attack plans and mission details were briefed the next day by *General* Martin Harlinghausen himself.

One of the goals of this mission was the railroad near Sidi Bou Baker (approximately 60 kilometers NNW from Gafsa, see map on the following page.) For this mission three DFS 230's where equipped with three sabotage troops to be able to destroy the important railroad bridges. In the night of 27. December 1942 the first mission started and proceeded to its destined goal. This aerial caravan flew towards Cap Bon, and once they reached the sea they flew fly towards the South towards Sousse and Sfax.

During this flight the Ju 52/DFS230 combination met a returning group of Ju 88s, and one of the towing planes mistakenly followed the Ju 88 group. Due to this mishap the attack force was reduced to only two DFS planes. These two planes made it towards their goal and landed successfully at about 02:00 hours near the target bridge, which they destroyed. The command was then able to make it back to its own lines.

Another target was a bridge northeast from the town of Kasserine. For this mission three DFS230s were deployed and accordingly equipped. This mission was not as successful as the previous one. This attack group also lost a plane before reaching the target destination. However, this plane landed safely in the desert and the crew and the passengers were rescued one day later by German troops.

Still another DFS 230 went down approximately twenty kilometers away from the target. Unfortunately, this landing was not as soft as the other one and the glider pilot broke both his legs. An additional two soldiers in the plane were also wounded. The remaining unwounded continued with the mission, but all except two soldiers were captured by the French forces in the area. So this part of the mission clearly failed and did not achieve its goal of destroying the targeted bridge.

The picture below shows the area of operation with the targets.

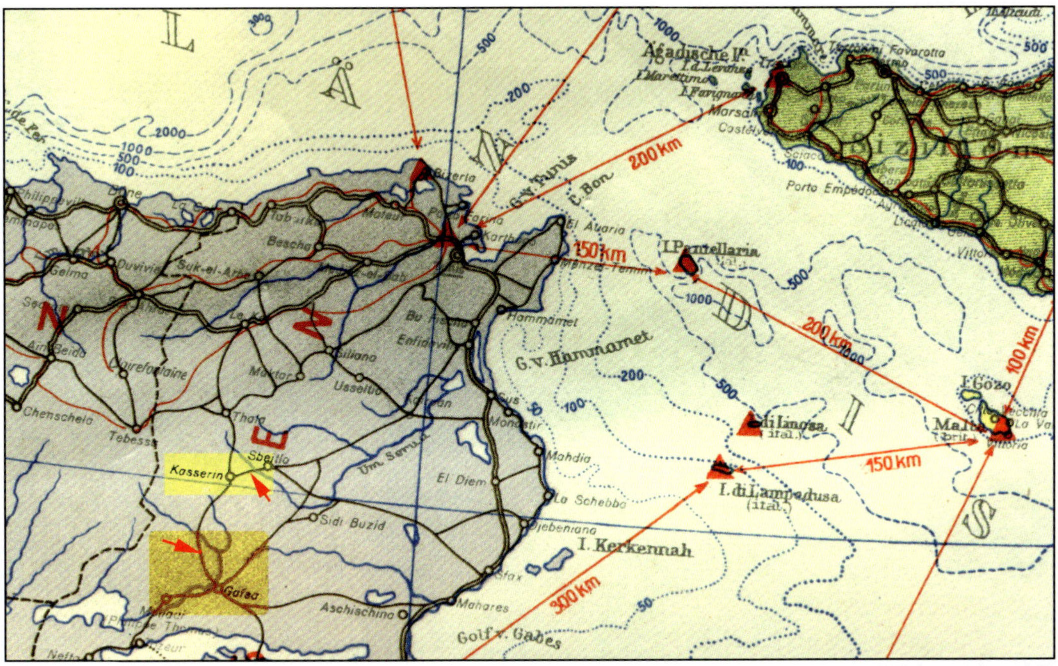


IRO Kriegs-Atlas, 1940

# Vassieux (France 1944)

The Vercors is a light mountainous area in the southeast part of France. The area around the town of Vassieux itself is a relative plain area in the middle of forests and woods. It was near that town that the French resistance arranged itself during the winter of 1942/43 with the goal of strengthening its forces against the German occupation forces.

Once the troops of the allied invasion in Normandy came closer to the resistance forces, those forces were reinforced up to a size of approximately 4000 troops by mid summer of 1944.

IRO Kriegs-Atlas, 1940

According to the Allied plans this area had a large importance since the Allied forces would need to pierce through it once they had landed in Southern France (Operation Dragoon). German supply lines which ran through the area were also vital to support any German counter attacks against the landing troops.

Very soon after the landings in Normandy on 6. June 1944 the resistance forces became more and more active, and each action of the French troops was followed by a reaction from the German side. By mid July 1944 the German forces were ready to attack and eliminate the partisans located in the Vercors area. To conduct this operation the following German units were assembled:

- Two Grenadier battalions and two Pioneer companies (both belonging to the 157. *Reserve Division*)
- *Polizei-Regiment* 19 (police regiment)
- *Sicherungsregiment* 200 (in parts)
- 3 East-Battalions
- *Feldgendarmerie* plus several parts of other units (military police)

A ground attack force had to pierce into the heart of the Partisan area. The following units had this task:

- Four *Gebirgsjägerbatallione* and two *Gebirgsbatterien* (belonging to the 157. Reserve Division - mountain troops)
- One battalion belonging to the *9. Panzer Division*

And finally an aerial assault force was assembled which was to land two companies of paratroopers through glider landings on the Vassieux plateau (where the partisan HQ was suspected of being located). Between 18. and 20. July several Do 17 - DFS230 towing combinations (belonging to the I./LLG 1) arrived at Lyon/France. These towing combinations were supposed to land the paratroopers.

In Lyon the operational briefing took place with aerial pictures. The gliders had to use the so-called "*Bremsfallschirme*" (brake parachutes) and stop rockets to be able to land successfully. In the morning of 21. July 1944 two groups of towing combinations started for the actual mission. The gliders of one glider group had to land near the villages of Le Chateau and La Mure. The second glider group had to land near Vassieux itself and capture these villages and destroy the enemy located there.

Approximately 10 kilometers before reaching the target the gliders where released at an altitude of 2500 meters (8200 ft.) and they reached the target after gliding through the air for about 5 to 8 minutes. At an altitude of 800 meters the *Bremsfallschirm* was deployed and the gliders went into a diving flight. Two gliders of the first group, which was attacking Le Chateau and La Mure, failed to deploy the *Bremsfallschirm* but still managed to land successfully.

The glider group with Vassieux as the target received very heavy ground fire which succeeded to hit 2-3 gliders very hard even before they could land. Through this ground fire two glider pilots were killed, as were a couple of paratroopers. Due to these losses the unit which was ordered to capture Vassieux was already weakened, and in the ensuing close combat the German assault force soon found itself in a very bad position. To help their comrades the force attacking Le Chateau and La Mure sent reinforcements. Although that assault force managed to capture both villages, they also started to feel that the Partisans were regrouping and starting to counterattack.

The result of this was that on the evening of 21 July 1944 both assault forces where decimated and in both cases they had become the defenders rather then the attackers. Losses on the German side where 29 KIA and 20 seriously wounded soldiers.

The next day did not bring relief due to bad weather, however the French resistance reinforced its forces even further. On July 23 better weather arrived and from the airfield in Valence another twenty DFS230 gliders took off with their towing planings, loaded with paratroops and *Ostlegionäre*. This force also experienced losses when one glider crashed during its landing and another one hit a rock formation, which resulted in the loss of both crews. Another glider had to land in the occupied partisan area but the crew was lucky enough to be rescued by a German infantry unit.

Together with the DFS230 gliders from the I./LLG 1 two Go 242 from the I./LLG 2) were deployed in the operation to resupply the troops. However, one cargo load of one of the Go 242s could not be retrieved by the troops. But with these new reinforcements the German defenders could once again take the initiative and hold out against the partisans. This was also possible due to the support from *Luftwaffe* ground attack air planes. In the area east of Le Chateau and Vassieux so-called several *Fieseler Storch* airplanes managed to land and take-off to evacuate wounded soldiers.

On the 24. July another Go 242 landed and brought in a 2 cm *Flak* gun. With the help of this gun several enemy strongholds could be destroyed. This was the turning point in the battle. The German attackers actually started driving the resistance forces back into the surrounding woods.

On 25.July 1944 the Germans started to withdraw the paratroopers and glider pilots into the direction of Grenoble. The German mountain troops consequently reached the villages, and so the battle died down. This operation, however, was undecided and both participants had serious losses.

The final numbers of casualties were not always clear. The losses on the French side were (according to one French source): 639 resistance soldiers and 201 inhabitants of the villages. German losses are not exactly known either, however certainly four glider pilots where KIA and several others were wounded.

Fieseler Storch (Stork), mainly reconnaissance plane

## Suchinitschi (Russia, 1941/1942)

At the end of September 1941 six Go 242 glider crews were drafted directly after their training and formed into a *Go-Einsatzkommando* (Go Task Force). The commando was first led by *Unteroffizier* Mirowsky (a former Crete glider veteran), who later handed over the command to *Oberfeldwebel* Held.

The Ju 52 – Go 242 glider combinations took off at Eschwege, West Germany to fly to Jüterborg, Posen and they arrived there on 2.October 1941 at their specific destination, the airfield of Smolensk/North, during very bad weather. The crews were placed directly under the command of the VIII *Fliegerkorps*. The first weeks did not require actual missions, and the usual waiting time started for the crews.

With the start of the Russian winter offensive this situation changed very quickly. The Russian forces pierced through the complete front of the *Heeresgruppe Mitte* (army group center) and continued to push westwards. Consequently German units were encircled, bypassed, and sometimes even destroyed while trying to form a defense line once again.

On 29.January 1942 strong Russian forces encircled parts of the 216th Infantry Division near Suchinitschi (a town approximately 230 km. southeast of Smolensk). The encircled German troops held an area of about 4.3 miles long by 2 miles wide.

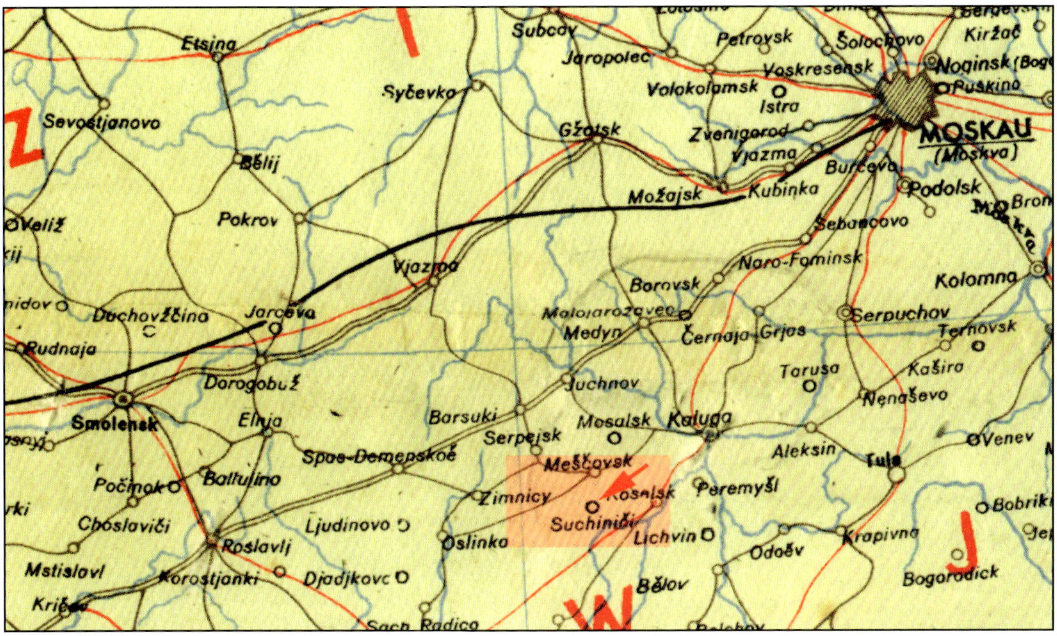


IRO Kriegs-Atlas, 1940

Inside this small perimeter nearly 5000 soldiers and another 1000 wounded were encircled.

According to received radio messages it became clear that the encircled German force lacked supplies such as artillery ammunition, weapons in general, food, and radio-transmitters. On 16. February 1942 the first Go 242 with crew and loaded with supplies headed towards the encircled troops. But this mission failed as the encircled area could not be located and safely identified as such. The Go 242 was towed back towards the German lines and landed safely.

The second Go 242 mission was more successful. The Go 242 glider was very heavily loaded, and barely managed to take off successfully. Once it reached the area over the encirclement the cloud cover was once again very tight so that the area could not be identified as such. However, the towing plane released the Go 242, so this glider had no other choice than to land. Green flare signals from the encircled troops pointed out where the German frontline was and the crew managed to land the Go 242 safely. Approximately half an hour later a second Go 242 also managed to land successfully.

The Russians had noticed this new method of flying in supplies and started very soon to guard the area with fighter planes in order to shoot down any new glider towing combos which attempted to fly in. But on 18. February 1942 two Go 242s were again towed in and the Russian efforts of interception were unsuccessful since their fighter planes were shot down by German fighters of the JG 54.

In total five Go 242 landed in the Suchinitschi encirclement and one other Go 242 went down in Russian territory before it reached the encircled German troops. One crew member of this latter aircraft made it back towards his own lines; the other one was MIA.

On 24. January 1942 the encirclement was broken when Suchinitschi was liberated by German tanks. The glider crews were brought back to Smolensk. This resupply mission was the first of its kind and was the start of a totally new way of use of the glider force. Officers from the formerly encircled 216th Infantry Division underlined the important role the glider resupply contributed to the resistance of the encircled troops. The previously utilized method of actually flying in resupplies or using drop containers also had its use, but the major difference with the use of the Go 242 was that this time large cargo loads of artillery ammunition and such could be brought in.

Amongst the missions that followed this first one were such successful employments of glider planes to supply the troops in the pockets of Cholm and Demjansk, just to name two.

## Welikije Luki (Russia, 1942/1943)

On 19. December 1942 a large Russian offensive started in the Southern front against the German *Heeresgruppe B* (Army Group B). At the same time German forces were attacked in the northern sector to prevent their transfer to the sections in danger at the southern front. The Russian goals of these local offensives were the strategic cities of Witebsk and Welikije Luki.

The town of Welikije Luki was of great importance as a crossing point of major roads and railroads. It was also exactly at the seam between the Army Group North and Army Group Center. Soon the city was encircled by Russian troops from the North and South and finally, on 26. December 1942, the German forces in that area were completely surrounded.

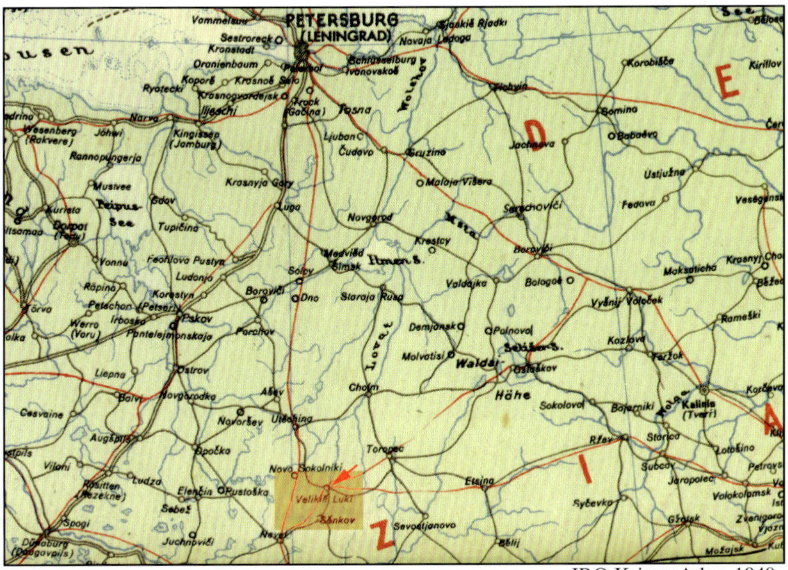

IRO Kriegs-Atlas, 1940

Immediately after the encirclement, the *Luftwaffe* formed a special command under the leadership of the *Kampfgeschwader* 53 and under the command of *Oberst* Wilke to help the encircled German troops. Several fighter, bomber, transport, and glider groups were attached to give the command the highest chance of success.

The largest problem of the encircled German troops was that they lacked sufficient supplies to be able to defend themselves effectively against the Russian attacks. The *Luftwaffe* immediately started with dropping resupply containers. However, this supply solution was not very useful as a complete solution. The container dropping air crafts had to fly in at very low altitude to make sure that the containers landed within the German perimeter. The obvious problem with this low altitude was that Russian anti-aircraft artillery soon guarded theses flying corridors and was determined to cut this aerial lifeline. Another problem was a very bad weather front moving through the area which prevented the *Luftwaffe* from effectively flying on several days.

Therefore another solution was - once again - found: flying in resupplies with glider planes. The landings had to be done at dusk or during the night and also near the railroad station of Welikije Luki. To make this possible the landing strip had to be lighted to allow the gliders to land safely. There was no personnel inside the cauldron who knew how to professionally conduct an airstrip lighting, so a volunteer was parachuted in to prepare the coming glider landings.

On 16.December 1942 the first tow plane/glider combination flew to the encircled area. The load of the Go 242 glider was three tons of medical supplies and *Feldpost* (army mail). The Go 242 found the air strip and landed successfully. However before the glider could stop it hit a house and the crew members were wounded during this accident. The cargo load was successfully unloaded and the first Go 242 supply mission to Welikije Luki was in the history books. The weather turned bad again so that no further glider landings could be made in the upcoming days.

The Russian forces however did not take a minute rest and kept the pressure at a high level, causing permanent German losses. These constant losses in combination with failing resupply started to paralyze the defending forces. German efforts to reach the encircled forces over ground were also unsuccessful.

On Christmas day 1942 the weather finally cleared up and another Go 242 glider was towed to the encirclement. The Go 242 was hit by enemy ground fire while flying in and crashed burning before it could reach the German lines. This crash resulted in the loss of the crew and the passengers (16 soldiers). The He 111 planes which flew drop container missions also lost one airplane over the city. The following days the weather turned bad and it was again impossible to fly supply missions.

However, on the 29. December 1942 five gliders landed safely in the morning, and an additional two planes landed safely at noon to bring in the needed supplies. On 30.December a German glider of the type DFS230 crashed at Welikije Luki, and on the last day of the year 1942 the *Luftwaffe* could fly in 4 Anti-Tank guns with its gliders. These small successes were overshadowed by the fact that the Russians cut the encircled German troops into two parts. Due to this fact the German defenders were weakened very much, since there was no longer only one front to fight. The two groups of defenders now each stood on their own. One part of the German defenders was located near the Railroad station (eastern part). The other part of the defenders was concentrated in and around the so called "Fortress."

The He 111s spared no cost and flew in over and over again dropping their containers. Due to the fact that the German defenders were no longer together in one area the majority of the dropped supplies did not land in German hands. Still the *Luftwaffe* did not give up and on the 6. January 1943 again two glider planes were again towed in and released to help the defenders. One of these gliders was shot down and crashed. The second glider landed safely.

Only very few glider planes were towed towards the eastern part of the Welikije Luki encirclement in the first part of January. On one day two Go 242s were towed to the encirclement and when they landed they where hit by AA fire. The gliders were able to land, but in one of the gliders the board mechanic was killed by this fire. The pilot of the glider reached the German lines, but he was then hit by enemy fire that killed him at the spot. The second glider was loaded with ammunition which was set on fire by the AA fire.

The efforts of the bomber- and glider crews resulted in the prolongation of the inevitable end and the encircled troops held out for a few more days. The defense position of the "Fortress" finally fell to the Russian on 16 January 1943 after vicious fighting (picture left). The eastern part also capitulated on this same day. The wounded German defenders were gathered and placed on a transport to a POW camp. The remaining soldiers had to walk for miles through the snow and many of them, who had survived the battle, collapsed from exhaustion. The soldiers who collapsed and were not able to walk anymore were shot on the spot by the Russian guards.

Out of the the nearly 2000 soldiers who surrendered, only between 600 and 800 reached the prison camp. The balance did not survive the death march.

A total of 17 gliders successfully landed in the Welikije Luki encirclement (11 Go 242s and six DFS 230s). Two German gliders were shot down and crashed on Russian territory before they reached the encircled troops and at least another six gliders were towed to the encirclement but could not be released and were towed back to the German lines. A total of 30 members of the glider troops where KIA or went MIA in this unsuccessful supply mission.

## Berlin (Germany, 1945)

At Köthen (Saxony/Anhalt) a small grouping of glider pilots was drafted together and sent to Wittstock/Dosse. Here the glider pilots had to prepare themselves to be ready for a special mission.

In the last week of April 1945 these glider pilots where transported to Tarnowitz via trucks. Here the glider planes were made ready for action and were flown to Rerik (at the Mecklenburg coast). The greater parts of Mecklenburg were already occupied by Russian forces.

The goal of this small glider troop was to fly into Berlin to resupply the city. This insane mission was ordered by Adolf Hitler in a final attempt to bring in fresh supplies and then defeat the Russians. Berlin had been encircled since the 25. April by the Russian forces. This mission was a typical example of how strangely German forces were used in an effort to make the desperate situation last a few days, or even weeks, longer.

The defense of Berlin was one of the most senseless battles in history. The Russian army had a force of approximately 6200 tanks, 41,000 artillery pieces, 8000 planes and 2.5 million soldiers at its disposal to capture the city. The German forces existed of approximately 90,000 soldiers, but only on paper. The soldiers were actually, for the largest part, drafted *Volkssturm* soldiers and *Hitlerjugend* boys and girls.

The Russians had plenty of ammunition, weapons and gasoline. The German defenders lacked supply in every category. The few soldiers with combat experience were mostly coming from destroyed units which were formerly deployed at the Oder front. The German capital still contained nearly three million civilians.

In this situation the gliders were ordered to fly in and bring the much needed supplies to help to defeat the Russian attackers. On the 27. April the last open airfield of Berlin (Gatow) was captured by the Russians, so no more incoming planes could land or fly out of the beleaguered city. Therefore the gliders had to land on a few large streets. This also was only possible during the night as flying in during the day was impossible due to Russian fighter planes and anti-aircraft guns. Only missions at night stood a slight chance.

In the night from the 28. April to the 29. April five glider planes (Type DFS 230) started from Rerik to the encircled capital. The designated landing spot was under such a heavy artillery barrage that to land there was equal to suicide. Therefore each glider had to find its own landing place. One of the glider pilots, *Unteroffizier* Kugler, oriented himself with the help of fires in the city and found a suitable spot near Tegel (a part of Berlin), and managed to land safely there. The ammunition cargo he had on board was immediately requisitioned by a nearby German patrol. This glider pilot survived the war but the fate of the other four gliders is not known and the pilots were MIA from that day on.

Every one of the glider pilots carried a special permission with them as they were seen as specialists who could be re-used. With this permission they where granted a priority if a plane managed to land in Berlin and was able to fly out again, taking some lucky persons to safety. And so *Unteroffizier* Kugler this glider pilot had the luck on his side and managed to catch one of the last departing transport planes which left Berlin to fly him out.

During the next few days another two DFS 230 gliders flew the mission to Berlin. One of the incoming gliders was greeted with heavy anti aircraft fire but landed safely. The pilot was heavily wounded and was immediately captured by the Russians. The other glider pilot did not survive the supply mission and was killed when he landed.

The remaining glider pilots at Rerik did not need to fly this absolutely crazy mission. They joined a transport group and flew out of Rerik to Eggebek (Schleswig-Holstein) and went into English captivity.

The transport mission to supply Berlin was the last German glider mission of WWII. The brief descriptions of the missions in this book give a fairly clear view of how the glider force transformed from a pure assault weapon at the beginning of the war to a unit which was even used in absolutely hopeless and desperate situations, simply to continue to fight a battle already lost.

An airplane wreck in the *Tiergarten* with the *Reichstag* in the background.

# THE INSTITUTION OF THE LUFTWAFFE GLIDER PILOT BADGE

The proposal to create a special Glider Pilot Badge originated from the 7. *Flieger Division,* and subsequently the *Reichsluftfahrtministerium* (RLM) asked the division to provide some concepts. One of the submitted ideas was created by the Berlin sculptor and designer Wilhelm Ernst Peekhaus, who received the design contract with the following guidelines:

> "The badges should show an eagle, which emblematizes the gliding and hovering of a glider airplane (i.e. para glider). The overall design should be similar to the other proficiency badges, such as the pilot badge."

*Generalfeldmarschall* Hermann Göring approved Peekhaus' design and it was given to the well known orders and medals manufacturer C.E. Juncker in Berlin with the assignment to create the badges.

The Luftwaffe Glider Pilot Badge was instituted on the 16. December 1940 by the State Secretary of Aviation and Supreme Commander of the air force, Hermann Göring. The institution, as well as the description and a drawing of the badge, was made public in the *Luftwaffe Verordnungsblatt, Nr.* 1635, 1940 (air force regulation no. 1635, 1940).

The document continues with the more detailed articles about this new proficiency badge:

**A. Description**

A solid and oval shaped oak leaf wreath made of metal. A swastika is located in the center of the lower edge of the wreath. On top of the upper half of the wreath a solid and oxidized eagle is attached by means of two rivets to the wreath. The reverse of the badge holds the brazed and oxidized needle and hinge.

Embroidered version: shape and form identical to the metal version. The 3-dimensional wreath is made with light aluminum, the eagle is made of oxidized embroidery and the swastika is manufactured with bare thread.

**B. Dimensions** (picture added for clarity)

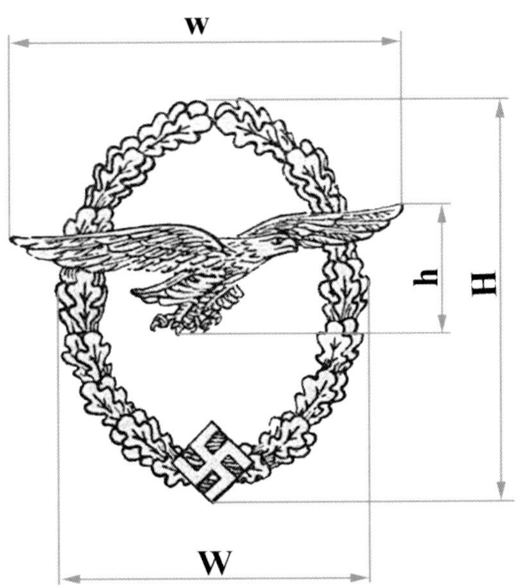


Dimension of the wreath:
H = approx. 55 mm
W = approx. 42 mm

Dimension of the eagle:
Wingspan w = approx. 53 mm
Height h = approx. 15 mm

**C. Material**

Badge is made of *Neusilber* (new silver or nickel silver)

## D. Award regulations of the *Luftwaffe* Glider Pilot Badge

1. Award qualifications

The LW-Glider Pilot Badge can be awarded to all active soldiers and soldiers on leave of absence with the required qualification and with the acquisition of the *Segelflugzeug-Führerschein (L)* [glider pilot license].

Civil servants, members of the engineering and navigational corps can acquire the LW-Glider Pilot Badge under the same conditions, provided they are soldiers on leave of absence.

On the occasion of the honorable discharge from the active duty of the flying staff of the *Luftwaffe* (including reserve duty) the flyers commemorative badge will be awarded if the leaving soldier was for a duration of at least four years in possession of the LW-Glider Pilot Badge or any other *Luftwaffe* qualification badge. An earlier awarding is possible on the occasion of a discharge due to a non-self inflicted accident. The decision rests with the *Reichsluftfahrtministerium, Personalamt* (Air Force Ministry, Personnel Office or RLM-LP).

2. Award application

Award applications for the LW-Glider Pilot Badge have to be signed either by the air born troop squadron commodore or by a commander of a glider pilot school of the Flieger Division 7, resp. their deputies. The application in two copies must be sent to the RLM-LP for a direct and accelerated decision.

Compliance to certain dates during the submission of the application is not required. The applications must use the standard form as per enclosure (see below).

**Anlage**
zu Nr. 1635

**Muster**

| Lfd. Nr. | Jetziger Dienstgrad (und Wehrverhältnis) | Vor- und Zuname | Geboren am | Dienststelle | Segelflzgf. Sch. (L) | | | Besitzt welches Abzeichen? Seit wann? |
|---|---|---|---|---|---|---|---|---|
| | | | | | Nr. | Datum | ausgestellt von | |
| | | | | | | | | |

The actual awarding takes place with the date of the (document) signature. Retroactive awards are not possible.
The badge will be sent to the requesting department together with the award documents for immediate handing over to the awardee. The date of the award needs to be noted into the *Wehrpass* in the case of the officers, clerks and members of the engineer and navigation corps. For NCO and regular troops the award will be recorded into the personnel roster.

The award is to be recorded into the dated inventory adjustments as part of the personnel files of the RLM (LP). Awards of the LW-Glider Pilots Badge made to soldiers on leave of absence are to be announced by the requesting department to the responsible military district command post with a short notice. If the award takes place after the soldier already reverted back to his civil occupation the award and the documents will be sent to the responsible military district commando for the handing over to the awardee and also for the proper recording in the *Wehrpass.*

3. Regulations regarding wear of the LW-Glider Pilot Badge

The LW Glider Pilot Badge, as a military proficiency badge, is only authorized to be worn with the uniform of the armed forces or with the uniform of the *Nationalsozialistisches Flieger Korps* (NSFK). The LW-Glider Pilot Badge is worn in the middle of the left breast side of the tunic and on the service blouse as well as on the evening dress at the same position. If orders are present (EK1, ..) below them.

The LW-Glider Pilot Badge can be worn side by side with any Great War flyer badge.

Wearing a LW-Glider Pilot Badge together with any other *Luftwaffe* proficiency badge (other than the Parachutist badge) is barred. However, it is up to the owner of any other *Luftwaffe* proficiency badge to request a LW-Glider Pilot Badge. After it is awarded, the other Luftwaffe proficiency badge reposes. The wear with civil clothing, even as a miniature, is forbidden (Exception: Flyers Commemorative Badge).

4. Regulations regarding disposition of the LW-Glider Pilot Badge

In the case of transfer or relocation to another branch of the Luftwaffe the Glider Pilot Badge may be worn until a different proficiency badge has been awarded. After this new award the Glider Pilot Badge must be returned to the RLM (LP).
The Glider Pilot Badge together with the document is to be returned to the RLM (LP) when the owner leaves the active Luftwaffe service or transfers to the non-flying personnel.

In the case of death due to an airplane crash death due to no fault of one's own the badge worn by the casualty can be given to the dependents if they wish so. For the rest the regulations as announced in LVBl. 1940, S.486, Nr.990 apply.

**E. General**

1. The badge will be supplied at no cost to the recipient.

2. If the badge gets lost, the recipient has to replace the badge at his own cost.

3. Additional pieces, also in embroidered form, can be bought at the *Luftwaffe* sales office (LVA or *Luftwaffe Verkaufsabteilung*), Berlin SW 68, Puttkamerstr. 16-18.
The LVA is instructed to sell such pieces only when the proper original documents or duly notarized documents are shown. If ordered via mail an attestation of the relevant department with the official seal needs to be included.

4. Any unauthorized wear of the badge while belonging to the armed forces will be punished by disciplinary measures.

5. In the case of disciplinary or court punishments because of breach of flight discipline or military insubordination (Paragraph 92 *MSt.GB*) in direct connection with a breach of flight discipline newly awarded badges will only be given out after an additional probation period. Earlier awarded badges are disallowed to wear for a given time set by the disciplinary superior with at least the disciplinary penal power of a squadron commander. The issue has to be reported to the *RLM* (LP).

6. Pricing of the badges

| | | |
|---|---|---|
| In metal ...................................... | ~ | 5.00 RM |
| In hand-embroidered form ......... | ~ | 7.50 RM |
| In machine-woven form ............. | ~ | 0.90 RM |

Der Reichsminister der Luftfahrt
und Oberbefehlshaber der Luftwaffe

i.V Milch
Generalfeldmarschall

# THE TRAINING UNITS OF THE GLIDER TROOPS

To understand the history of this badge it might be advisable to know a little about the training units and their structure.

The basic precondition to train as a Luftwaffe glider pilot, and therefore get the admission to one of the Luftwaffe glider training schools, was the possession of a civil glider license, class II. This meant that one had to successfully complete at least 40 hours of flying a glider (double seater).

Already in January 1940 a "Glider School of the *Luftwaffe*" under the command of *Hauptmann* Willerding at the airport Brunswick-Waggum (Lower Saxony) was established. After the successful completion of the campaign in the west a total of 60 soldiers (motorized personnel and glider teachers) as well as eight Ju-52 tow-airplanes and 20 gliders (type DFS 230) of the 17./KGzbV5 ( the unit which conducted the landings near the Albert Channel on May 10, 1940) were transferred to this new school.

This school was transferred in the late summer of the same year from Brunswick to Neuhausen (East Prussia). *Major* Ludwig Reeps took command of the now renamed "Flyer School of the Luftwaffe (S) 1." The school was again transferred, most likely on the 10. April 1941, to Langendiebach (Hessia) in western Germany. The transfer was a result of the upcoming invasion of Russia (the *Luftwaffe* needed the airport for its deployment).

Together with the overall restructuring of the glider troops in the spring of 1942 the "*Fliegerschule (S) 1*" (flyer school) was renamed to "*Ergänzungstruppe (S) 1*" or "Erg. Gr.(S))" [replenishment group (S) 1].

Beginning with the summer of 1943 the operations of the flyer schools inside of Germany were seriously hindered due to the ever deteriorating situation on the Russian front and the Mediterranean area. Most of all the ever decreasing gasoline resources resulted in a diminishing supply to the training units. Finally the *Ergänzungsgruppe (S) 1* was, together with all the other glider units, dissolved as per an order from the *Oberkommando der Luftwaffe* (Supreme Airforce Command or OKL) dated 9. September 1944.

Therefore no more Glider Pilot Badges were awarded after this date, as with the disbanding of the training units no more pilots were trained.

Parallel to the Erg.Gr.(S)1, another unit was formed, the "*Ergänzungsgruppe (S) 2.*" This unit which incorporated four glider squadrons , was formed out of the old "*Schleppschule der Luftwaffe*" (air force tow school) by renaming it to Erg.Gr.(S)2. As was the case with

*Erg.Gr.(S)1*, this school, located in Posen, was also tasked with the training of glider and tow plane pilots.

The end for this school, however, came fast and also very unexpectedly. To cover the retreat of the *Panzerarmee Afrika* (tank army Africa) a defensive line needed to be formed in Tunesia. Since it was obvious that there were no more reserve units inside of Germany, the Erg.Gr.(S)2 was regrouped and renamed into an infantery batallion (*I.Batallion/LW Feld Reg. Barenthin*) and transported to Tunesia in November 1942 to fight as ground troops.

The remainder of the battalion went into Allied captivity after the surrender of the German troops in Tunesia on 13. May 1943. (Only very few soldiers managed to escape to Italy). Meanwhile, the training courses in Posen were conducted by a residual command which served as a core group when the *Erg.Gr.(S) 2* was reformed in the summer 1943. Finally this school also received the same order to disband on the 9. September 1944 and the training ceased.

Parallel to the actual training schools, other *Luftwaffe* units, such as the LLG 1 and the *1.(Go)/KGzbV 1*, trained glider and tow plane pilots in significant numbers.

# ACQUIREMENT OF THE GLIDER PILOT LICENSE (L)

The glider pilot license (L) ("*Segelflugzeugführerschein (L)*") was introduced on 1. November 1940, which was about six months before the institution of the Glider Pilot Badge.

The glider pilot license (L) was issued - as per the regulations - to soldiers and officials of the *Luftwaffe.* It entitled the owners to operate a glider out of a dedicated airport area.

The regulations state in particular:

I. The owner of a glider pilot license (L) will be called "glider pilot." They belong to the flying personnel and are per 1. November 1940 under the supervision of the *Luftwaffe* personnel office (LP).

II. The glider pilot license (L) will be issued in the classes A and B.
Class A: para gliders seating 10 persons.
Class B: para gliders seating 20 or more persons.

III. Each candidate for the glider pilot badges (L), class A, has to show the following requisites:

a. the *Luftwaffe* glider license class II, or
b. the glider license class II (civil), or
c. the *Luftwaffe* airplane pilot license

IV. Each candidate has to pass a final exam, consisting of a theoretical and a practical part:

1. The theoretical part consists of the theoretical test A2 according to L.Dv. 21 (*Luftwaffen Dienstvorschrift* or service regulations) without sections about the engine, aircraft, and instruments, but concentrating on the par ticularities of the gliding.

2. The practical part comprises the following flights:
- three target landings, tow release at 500 meters (approach with side, rear and head wind), must land on a field of 50 x 100 meters.
- two target approaches with map (over 30 km, glide path), one flight at twilight.
- one formation flight with 6 planes with subsequent precision landing on several designated points, tow release at 1000 meters.

The practical part (continued):

- one target flight (over 30 km, gliding), landing on a terrain which is only known by map or photograph; followed by a tactical training in conjunction with glider troops.

- one tow-formation flight of 2 hours duration with six planes; followed by a target flight over 20 km gliding and precision landing.

- three night landings.

Remark: All flights are conducted with two pilots. The taking on board of additional passengers is only permitted for the terrain landing with subsequent tactical training and in general only after obtaining the glider pilot license (L).

V. The practical as well as the theoretical test to obtain the glider pilot license (L) can only be conducted at the flying formations respectively, at the glider schools of the Flyer Division 7. The issuance of the license can only be done by the flying formations respectively, schools of Flyer Division 7.

VI. The duration of validity of the glider pilot license (L), class A and B, is one year, starting with the date of issue. To retain the license it is mandatory to fulfill the minimum requirements during the duration of one year: total of at least 15 flying hours (tow or gliding) with a para glider.

VII. The borders of physical fitness are determined by the "regulations for the medical examination regarding fitness to serve as a pilot, parachutist and parachutist during war." The candidate for the glider license needs to be fit to serve as a pilot.

VIII. Flyer Division 7 is responsible for any confiscation of glider licenses (L).

IX. ......

X. ......

Enclosures

Distribution list

Before a candidate could even start with the practical part of the test he had to conduct all the basic requirements of the flyer education as shown in the appendix.

After the training with the glider the flyers were put through an infantry base course. At the end of each course approximately 65 to 70% of the initial participants made the grade and received the glider pilot license (L).

All the flights conducted during the practical part, i.e., the training with the para glider type DFS230, were noted in the glider flight book (*Segelflugbuch*). The flights were confirmed by the course teacher in due time and served as evidence that the required flights had been conducted successfully.

The basic requirements to obtain the *Luftwaffe* glider license (*Segelfliegerschein*) or the glider license for glider pilots (*Luftfahrerschein für Segelflugzeugführer*) were the same. The only difference was that one was issued by the armed forces, i.e. the *Luftwaffe* and the other was issued by civil authorities. This is the reason that either was accepted as a requisite for the glider pilot license.

Both types - the *Segelfliegerschein* and the *Luftfahrerschein* - are discussed in more detail at the end of the book.

As a summary, here are the steps that had to be followed and successfully concluded to obtain the *Luftwaffe* Glider Pilot Badge:

1. Possession of the glider license class II
2. Successful conclusion of the glider training to obtain the glider pilot license (L)
3. Application to the RLM, issued by an authorized person or department.
4. The document and the badge is issued upon approval and sent to the requesting unit for the award procedure.

# THE GLIDER PILOT LICENSE (L)

As of the date of this book the author is aware of five units or troops which issued the glider pilot license (L). Those were:

- *Luftwaffe* Flyer School (S) 1, resp. Erg.Gr.(S)1
- *Luftwaffe* Flyer School (S) 2, resp. Erg.Gr. (S) 2
- *Kampfgeschwader zur besonderen Verfügung 1* , resp. KGzbV 1
- I. *Gruppe Luftlandegeschwader 1*, resp. I/ LLG 1
- III. *Gruppe Luftlandegeschwader 1*, resp. III/LLG 1
- *Stab Luftlandegschwader 2*, resp. Stab/LLG 2

The picture (above) shows the glider pilot license (L) No. 50, issued to the private Paul Hagen by the *Luftwaffe* Flyer school in Neuhausen/East Prussia and dated 1. May 1941.

For the more interested reader a table follows which shows the dates in connection with the license numbers. It is known that every issuing unit used its own numbering system. The appendix of the book shows some of the very early glider pilot licenses and the corresponding units.

| No. | Date | Name | Rank | Unit |
|---|---|---|---|---|
| 3 | 24. Dec. 40 | Opitz, Rudi | Feldwebel | LW Flyer School (S) 1 |
| 50 | 1. May 41 | Hagen, Paul | Obergefr. | LW Flyer School (S) 1 |
| 66 | 1. May 41 | Sauer, Karl | Oberlt. | LW Flyer School (S) 1 |
| 93 | 15.May 41 | Becher, Heinz | Gefreiter | Erg.Gr. (S) 1 |
| 140 | 1. July 41 | Oberle, Günther | Flieger | Erg.Gr. (S) 1 |
| 143 | 25. Aug. 41 | Reichert, Adam | Feldwebel | Erg.Gr. (S) 1 |
| 232 | 15. Aug. 41 | Schaedler, Wilhelm | Gefreiter | Erg.Gr. (S) 1 |
| 240 | 26. Aug. 41 | Bethke, Kurt | Flieger | Erg.Gr. (S) 1 |
| 288 | 1. Sept. 41 | Grimm, Wolfgang | Flieger | Erg.Gr. (S) 1 |
| 312 | 15. Sept. 41 | Gerstmann, Karl-Heinz | Flieger | Erg.Gr. (S) 1 |
| 463 | 14. Jan. 42 | Rocktäschel, Erich | Flieger | Erg.Gr. (S) 1 |
| 467 | 14. Jan. 42 | Stark, Helmut | n/a | Erg.Gr. (S) 1 |
| 544 | 15. April 42 | Link, Joseph | Flieger | Erg.Gr. (S) 1 |
| 857 | 15. April 44 | Bortfeld, Jürgen | Gefreiter | Erg.Gr. (S) 1 |
| 24 | 25. July 42 | Kahm, Franz | Gefreiter | Erg.Gr. (S) 2 |
| 84 | 3. May 43 | Stangl, Franz | Gefreiter | Erg.Gr. (S) 2 |
| 4 | 26. Juni 41 | Röhrle, Anton | Obergefr. | 1./KGzbV. 1 |
| 34 | 10. July 41 | Auritsch, Josef | Gefreiter | 1./KGzbV. 1 |
| 44 | 1. Febr. 41 | Baier, Siegfried | Obergefr. | 1./KGzbV. 1 |
| 88 | 1. Dec. 40 | Schulz, Helmuth | Flieger | Stab I/LLG 1 |
| 121 | 15. Sept. 41 | Sobrig, Emil | Flieger | Stab I/LLG 1 |
| 131 | 16. Oct. 41 | Schneider, Karl-Heinz | Flieger | Stab I/LLG 1 |
| 24 | 1. Febr. 41 | Waldner, Siegfried | Flieger | III/LLG 1 |
| 96 | 26. Sept. 41 | Bracht, Otto | Flieger | III/LLG 1 |
| 3 | 30. June 42 | Ullrich, Erich | Flieger | Stab/LLG 2 |
| 6 | 15. July 42 | Dumke, Walter | Leutnant | Stab/LLG 2 |
| 22 | 1. Aug. 42 | Kremmer, Hans | Gefreiter | I/LLG 2 |

The earliest date for a glider pilot license (L) that is known to the author is dated 1. December 1940. The entrance in the service record is dated 26. February 1941. This might be an indication of the processing time to prepare the award document and send out the badge itself, which was about two months in this case.

It is also interesting to note that relatively few licenses were granted in the time frame between December 1940 and mid-1941. After that the amount increases very quickly. The explanation for this could be that the training itself went very smoothly, and the delay was actually in the processing and handing out the documents.

It has been already mentioned earlier that only owners of the glider pilot license (L) could apply for the LW-Glider Pilot Badge. The license was an absolute prerequisite to receive this rare badge and the accompanying document. This will be discussed in later chapters.

It is possible by means of the issued licenses to determine the approximate number of trained pilots. Presumably this number was approximately 1700 para glider pilots. And the number can be divided amongst the schools as follows:

| | | |
|---|---|---|
| * Erg.Gr. (S) 1: | | 900 |
| * Erg.Gr. (S) 2: | | 400 |
| * others such as KGzbV1, LLG1, LLG2, .. | | 400 |
| | | ===== |
| | Total | 1700 |

The numbers given for any unit (e.g. ErgGr. (S) 1) are for the time frame between the first award date and the termination of the para glider unit on 9. September 1944. The total is only an approximation and should not be taken as final number, meaning that ongoing investigations might very well correct the number.

## A DETAILED LOOK AT THE GLIDER PILOT LICENSE (L)

The license booklet has a red/orange cover and the front shows the *Luftwaffe* eagle. Underneath is the title of the license: *Segelflugzeugführerschein* (L).

The first page shows the license number, a picture of the owner, his signature, and the issuing department. The license always bears three agency stamps.

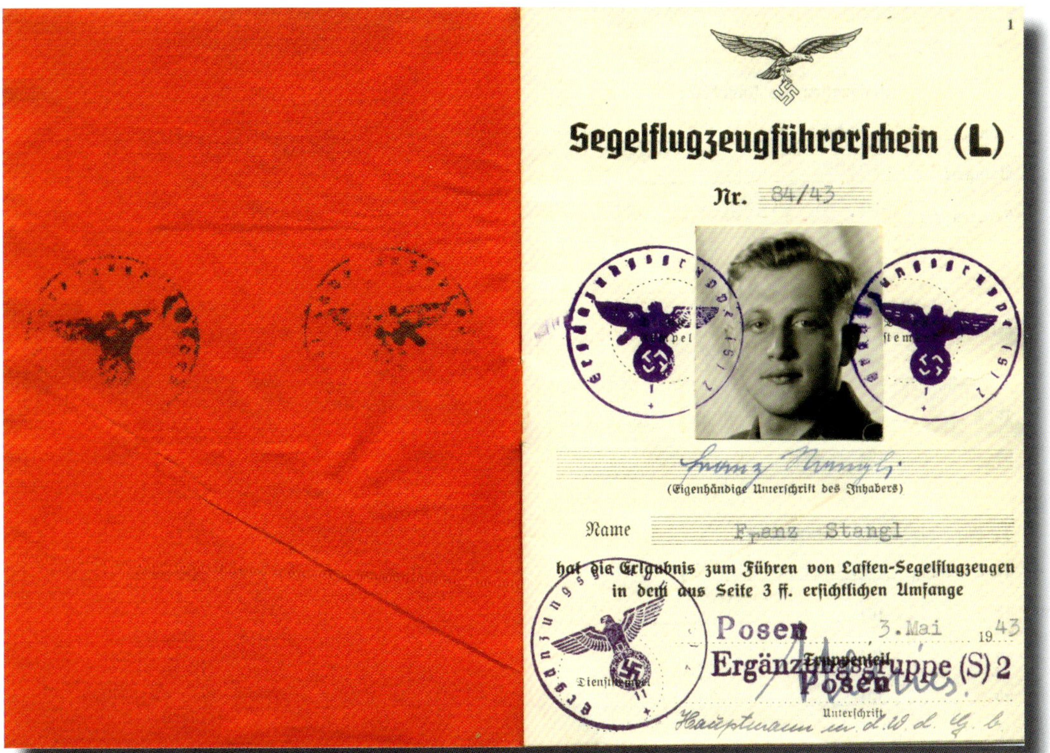

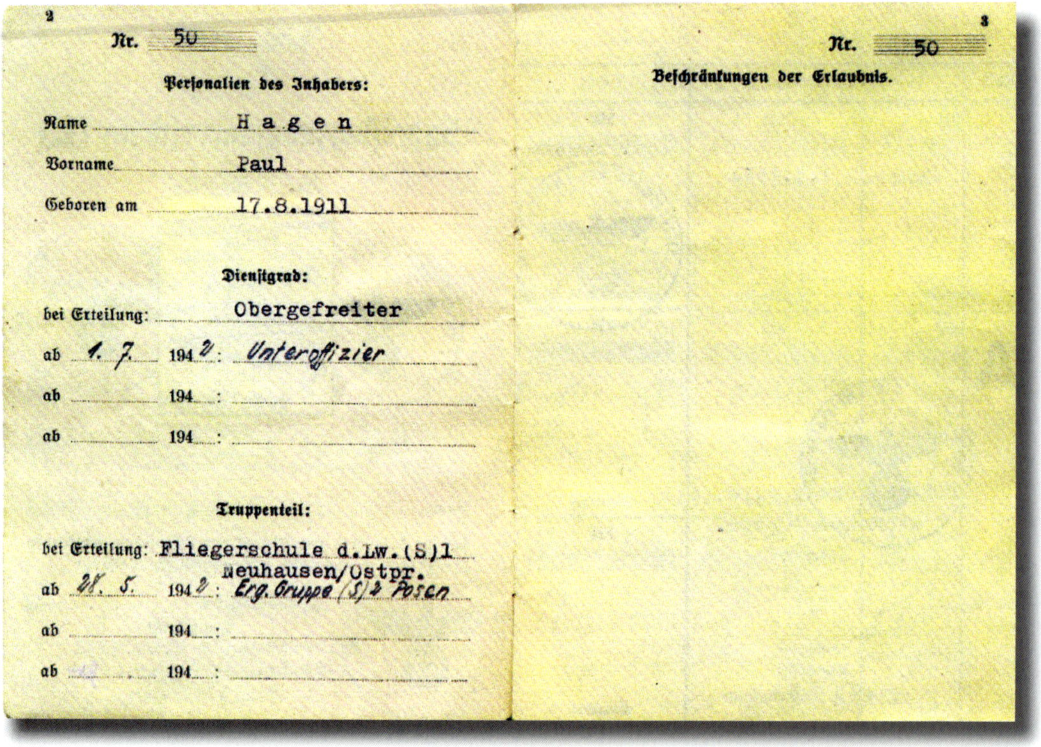
2

Nr. 50

**Personalien des Inhabers:**

Name Hagen

Vorname Paul

Geboren am 17.8.1911

**Dienstgrad:**

bei Erteilung: Obergefreiter

ab 1. 7. 1942: Unteroffizier

ab ...... 194 .. : ......

ab ...... 194 .. : ......

**Truppenteil:**

bei Erteilung: Fliegerschule d.Lw.(S)1 Neuhausen/Ostpr.

ab 28. 5. 1942: Erg.Gruppe (S)2 Posen

ab ...... 194 .. : ......

ab ...... 194 .. : ......

3

Nr. 50

**Beschränkungen der Erlaubnis.**

Page number two held the personal data of the owner as well as the service grade and the units in which he served. The following line items were recorded:

- Name
- First name
- Date of birth

- Service grade when license was issued
- Any possible promotions

- Unit served when the license was issued
- Any other units transferred to

On page three there was ample room to note any restrictions of the license itself. Up to this date the author has not found one glider pilot license in which any such restrictions were noted.

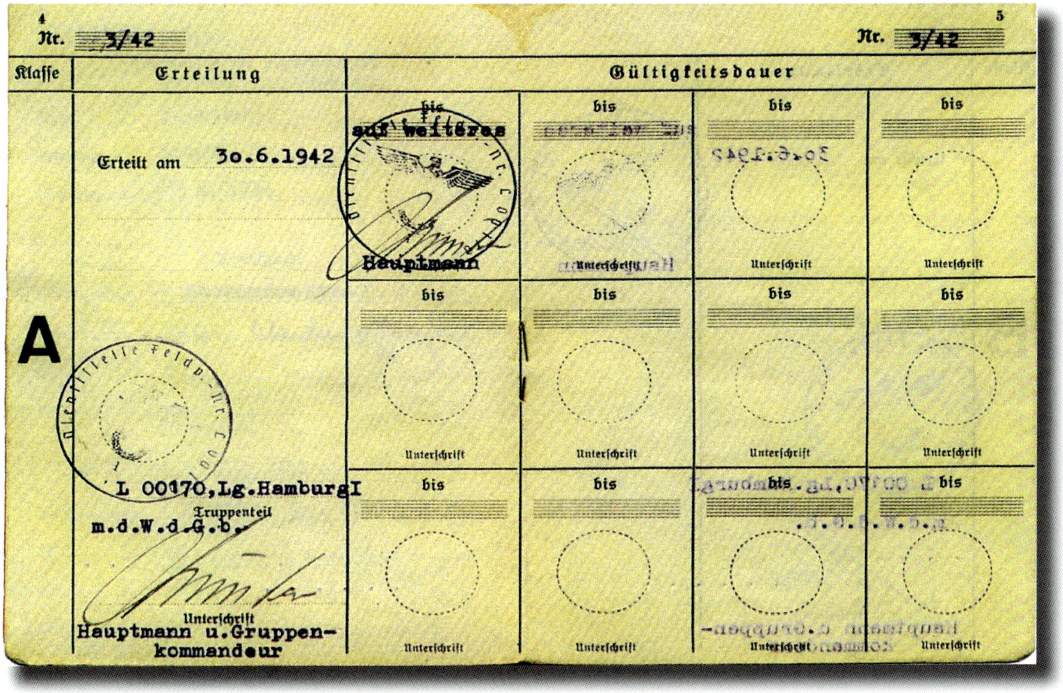

4 Nr. 3/42 — 5 Nr. 3/42

| Klasse | Erteilung | Gültigkeitsdauer | | | |
|---|---|---|---|---|---|
| A | Erteilt am 30.6.1942 | bis auf weiteres<br>Hauptmann | bis<br>Unterschrift | bis<br>Unterschrift | bis<br>Unterschrift |
| | | bis<br>Unterschrift | bis<br>Unterschrift | bis<br>Unterschrift | bis<br>Unterschrift |
| | L 00170, Lg. Hamburg I<br>Truppenteil<br>m.d.W.d.G.b.<br>Unterschrift<br>Hauptmann u. Gruppen-kommandeur | bis<br>Unterschrift | bis<br>Unterschrift | bis<br>Unterschrift | bis<br>Unterschrift |

The extreme left column (above) denominates the class (*Klasse*) of the license. In this case the capital "A" means that the owner was certified to fly military glider planes up to a capacity of 10 passengers, i.e., such a plane as the picture below shows.

The second column shows the date of granting the class permission together with the stamp and signature of the issuing troop unit, in this case the L/00170, which is the LLG2.

The normal validity period of the license was one year and the minimum requirements to uphold the license were so easy to achieve by the pilots that it was a very rare occurrence if a soldier did not pass them (see also page 20).

The picture shows a glider type DSF 230 with the markings S5 1-5. This plane type was used for the attacks on the Belgian border fort Eben Emael, at Crete and during the Mussolini mission at the Grand Sasso, among others. It was also used for strictly supply missions, as the one at the Kuban toehold. During WW2 about 1020 gliders of this type were manufactured.

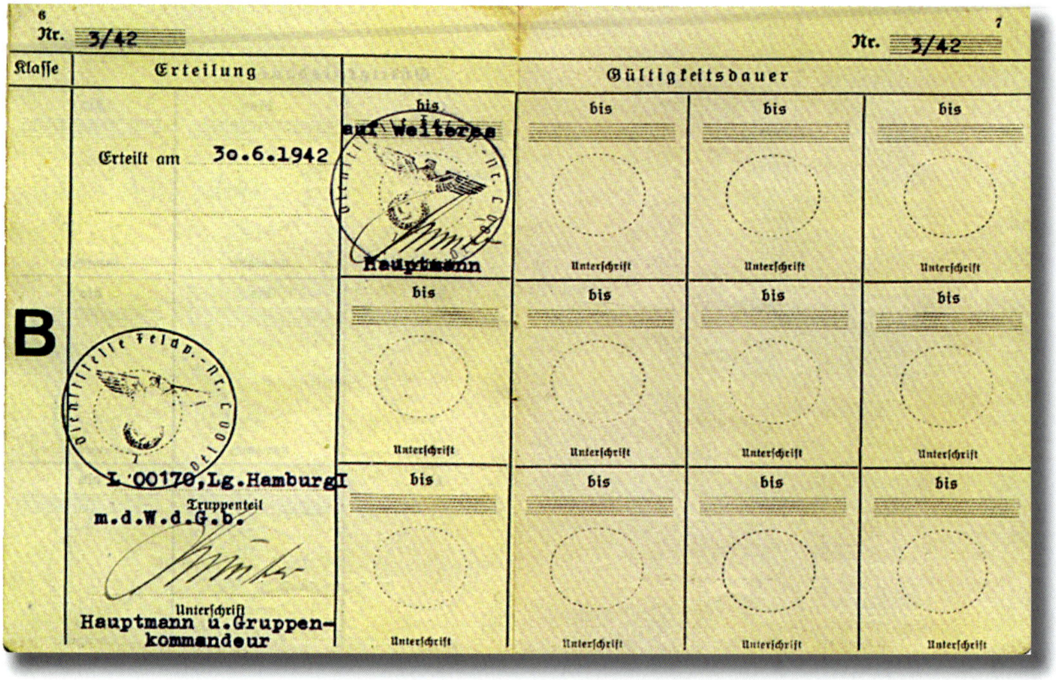

6 Nr. 3/42

7 Nr. 3/42

| Klasse | Erteilung | Gültigkeitsdauer | | | |
|---|---|---|---|---|---|
| B | Erteilt am 30.6.1942 | bis auf Weiteres<br>Hauptmann | bis<br>Unterschrift | bis<br>Unterschrift | bis<br>Unterschrift |
| | | bis<br>Unterschrift | bis<br>Unterschrift | bis<br>Unterschrift | bis<br>Unterschrift |
| | L 00170, Lg.Hamburg I<br>Truppenteil<br>m.d.W.d.G.b.<br>Unterschrift<br>Hauptmann u. Gruppen-kommandeur | bis<br>Unterschrift | bis<br>Unterschrift | bis<br>Unterschrift | bis<br>Unterschrift |

The class "B" as shown on pages six and seven of the license gave the owner permission to operate gliders with a capacity up to 20 passengers (together with the crew). As was the case with the previous two pages of the license, also this spread was also stamped and signed by the issuing troop unit.

Both class permissions, A and B, show the same date on this particular license. The normal case is that the dates of the issuing of the class A and B deviate from each other considerably. The reason for this is that each glider pilot had to go through the hierarchies of the training which meant that he was first schooled on the DFS 230 and only later on he was trained on the larger para glider Go 242.

The reason why this license has the same dates for both classes is easy. The owner lost his original licence during a mission at the eastern front and this particular license was reissued to him on the date for the second qualification.

The next page shows the para glider Go 242. This particular type was developed by the *Gothaer Waggonfabrik AG* (Gothaer Freight Car Factory Corporation). This company, already famous for its twin-engined bomber planes manufactured during the Great War, was located in the town of Gotha, Thuringia and was founded in 1898 to build railroad cars. Only in 1913 was an *Abteilung Flugzeugbau* (airplane department) added.

Roughly 27 years later, and after the initial successes of the gliders during the attack on France, the *RLM* decided to develop transport gliders with larger capacity. It was *Diplom Ingenieur* Albert Kalkert who came up with this new and innovative design. The underlying idea was to have a large and uninterrupted cargo hold with direct access at ground level. Therefore he opted for a shoulder-mounted wing arrangement. The twin boom tail was also arranged high above the fuselage and allowed for the hinged tail section to open up unobstructed.

This innovative glider had a capacity of 20 fully equipped soldiers plus the two pilots and a plane mechanic or a regular load of 2500 kg ( 5500 lbs). A total of 1528 Gotha Go 242 were built during the Second World War. They were mainly used as transporters and saw action at nearly all fronts. Multiple reports that one could not do without them are a testament to their usefulness. The glider was also used to transport bulky goods such as a *VW-Kübelwagen*, anti-tank guns, 2 cm *Flak*, spare parts for airplanes, and general supplies. But of course it was also used to transport regular troops and to evacuate wounded soldiers.

The picture (right) shows some Gothas Go 242s in the air at the eastern front. The camouflage paint scheme is very interesting.

# THE DOCUMENTS OF THE LUFTWAFFE GLIDER PILOT BADGE

The documents of the LW Glider Pilot Badge can be found in two different sizes classified by their DIN (*Deutsches Institut für Normung* or German Institution for Standardization, founded in 1917) format :

- DIN A 4 (210 mm x 298 mm , approximately 8" x 12")
- DIN A 5 (150 mm x 210 mm, approximately 6" x 8")

To make it easier for the collector these document types will be referred to as:

- Format DIN A 4 = Type 1
- Format DIN A 5 = Type 2

A document for the LW-Glider Pilot Badge cannot have an issue date earlier than December 16, 1940 (date of institution), and not after the end of 1944 (since the glider troops were disbanded on September 9, 1944 and one should allow a couple of months for an assumed processing time). It should be mentioned that the documents, depending on the processing time at the RLM, are in most cases dated later than the associated glider pilot license (L).

The listing of the known dates on the original documents makes it clear that the awards were always issued in batches to a certain amount of glider pilots. This is because the training was also done in groups. After the successful completion of the pilots training the newly issued documents together with the personal information were recorded in a list of awarded LW-Glider Pilot Badges.

The award regulations state (in extracts): "The award has always to occur with the date of the signature; retroactive awards are not possible."

All awarded medals (and documents) issued on a specific day were noted under that specific date in the awards list. As an example for this procedure one can use the entries for the 29. August, 1941:

| Datum | Number | Name | Rank | Size |
|---|---|---|---|---|
| 29. August 1941 | 85900/41 | Otto Schäuble | Flieger | A4 |
| 29. August 1941 | 85914/41 | Willy Jauch | Gefreiter | A4 |
| 29. August 1941 | 85930/41 | Gustav Gronlid | Obergefreiter | A4 |
| 29. August 1941 | 85931/41 | Siegfried Baier | Gefreiter | A4 |

The date on the award list is identical with the date on the proposal list, signed by the group commander or the head of the school. This list shows that there were at least 32 awards of the LW Glider Pilot Badge given out on August 29, 1941. The processing was handled in the usual bureaucratic manner, and the known dates were added to the proposal list.

Another example for this procedural method is revealed in chart 9, page 28 ( refer to the dates of September 5, 1941/ December 5, 1941/August, 4 1942/ April 20, 1942).

Another example of proof for the above outlined procedure is shown by taking a closer look at the three para glider pilots Schneider, Schurr and Martin. These soldiers belonged to the same pilot group (training for para glider pilot), and the successive numbers on the documents give further proof for list processing of the award of LW GPB (table below).

| Datum | Number | Name | Rank | Size |
|---|---|---|---|---|
| 26. November 1941 | 93220/41 | Karl-Heinz Schneider | Gefreiter | A5 |
| 26. November 1941 | 93221/41 | Hans Schurr | Flieger | A5 |
| 26. November 1941 | 93259/41 | Erich Martin | Flieger | A5 |

The picture below shows a group of trainees, among them Karl-Heinz Schneider (second from left) and Hans Schurr (left).

The names of both pilots in the picture (previous page) are also listed on the same application form " *Verleihung des LW-Segelflugzeugführerabzeichens*" (Awards of the LW Glider Pilot Badge). Both documents (see pages 41 and 42) were issued on the same day and the numbering is consistent.

The following chart lists all the award documents with date, document size, etc., as they are known to the author.

| Datum | Number | Name | Rank | Size |
|---|---|---|---|---|
| 23. January 1941 | 67034/41 | Ludwig Egner | Unteroffz. | A4 |
| 26. February 1941 | 70133/41 | Helmuth Schutz | Flieger | A4 |
| 26. February 1941 | 70172/41 | Heinz Powilleit | Uteroffz. | A$ |
| 12. April 1941 | 74308/41 | Werner Sinhart | Ogefr. (d.Res.) | A4 |
| 29. May 1941 | 77943/41 | Heinrich Neuhaus | Gefr. (d. Res.) | A4 |
| 5. Juni 1941 | 80362/41 | Anton Röhrle | Ogefr. | A4 |
| 29. July 1941 | 83161/41 | Karl Sauer | Olt. | A4 |
| 5. August 1941 | 83778/41 | Bruno Horn | Gefr. | A4 |
| 20. August 1941 | 84878/41 | Ullrich Erich | Flieger | A4 |
| 29. August 1941 | 85900/41 | Otto Schäuble | Flieger | A4 |
| 29. August 1941 | 85914/41 | Willy Jauch | Gefr. | A4 |
| 29. August 1941 | 85930/41 | Gustav Gronlid | Ogefr. | A4 |
| 29. August 1941 | 85931/41 | Siegfried Baier | Gefr. | A4 |
| 5. September 1941 | 86369/41 | Wilhelm Schädler | Gefr. | A4 |
| 5. September 1941 | 86389/41 | Adam Reichert | Fw. | A4 |
| 30. September 1941 | 88189/41 | Karl-Heinz Gerstmann | Gefr. | n/a |
| 4. October 1941 | 88547/41 | Wolfgang Grimm | Fl. | A5 |
| 26. November 1941 | 93220/41 | Karl-Heinz Schneider | Gefr. | A5 |
| 26. November 1941 | 93221/41 | Hans Schurr | Fl. | A5 |
| 26. November 1941 | 93259/41 | Erich Martin | Fl. | A5 |
| 5. December 1941 | 93603/41 | Gunther Oberle | Gefr. | A5 |
| 5. December 1941 | 93725/41 | Otto Bracht | Gefr. | A5 |
| 12. December 1941 | 94409/41 | Hans Mattes | Gefr. | A5 |
| 23. January 1942 | 96610/42 | Erich Rocktäschel | Fl. | A5 |
| 20. April 1942 | 101317/42 | Joseph Link | Fl. | A5 |
| 20. April 1942 | 101322/42 | Erwin Zahn | Ofw. | A5 |
| 12. May 1942 | 102238/42 | Alfred Wienröder | n/a | A5 |
| 4. August 1942 | 106258/42 | Franz Kahm | Gefr. | A5 |
| 4. August 1942 | 106276/42 | August Geiselauder | Gefr. | A5 |
| 7. August 1942 | 106384/42 | Klaus Bernett | Lt. | A5 |
| 24. November 1942 | 111170/42 | Rudolf Gotze | n/a | A5 |
| 6. March 1943 | 111875/43 | Josef Herres | Gefr. | A5 |
| 6. March 1943 | 123756/43 | Franz Stangl | Gefr. | A5 |
| 25. June 1943 | 127624/43 | August Schmidt | Gefr. | A5 |
| 9. September 1943 | 137854/43 | Heinz Heckerodt | n/a | A5 |
| 17. October 1943 | 142311/43 | Heinz Rieck | Gefr. | A5 |
| 29. April 1944 | 163108/44 | Walter Stohrer | Uffz. | A5 |

The transition from the larger A4 type document to the smaller (half the size A5) clearly happened between the 5. September and the 4. October 1941. It could very well have been that in the months following October 1941 both sizes were given out until the supply of blank A4 documents was exhausted. But there is no firm evidence for this assumption.

The document of Ludwig Egner, dated 23. January 1941, testifies (as of today) the earliest known award. Ludwig Egner already took part in the successful glider landing on top of the Belgian fort Eben Emael, and was already acknowledged before the war as an outstanding glider pilot.

The latest known document known as of today dates from the 29. April 1944. But it could very well be that later award dates exist since this was theoretically possible until the 9. September 1944.

All documents bear the signature of the department head of the *Luftwaffenpersonalamt* (LP or Airforce Personnel Office), either as an original or as a facsimile stamp. The signers were:

- *General der Flieger* Kastner-Kirdorf (from 1. September 1939 - 23. February 1943)
- *Generaloberst* Bruno Loerzer (from 23. February 1944 - 20. December 1944)

Gustav Kasner-Kirdorf signed either with the rank of *Generalleutnant* or as *General der Flieger* (from 1. July 1941 on). Bruno Loerzer either signed as *General der Flieger* or as *Generaloberst* (from 16. February 1943 on).

The time frame between the date of granting the glider pilot license (L) and receiving the document for the Glider Pilot Badge varies between several days up to several months. The following examples illustrate that point:

| | | |
|---|---|---|
| **Siegfried Baier:** | Glider Pilot License (L): | 1. February 1941 |
| | Document for the badge: | 29. August 1941 |
| **Adam Reichert:** | Glider Pilot License (L): | 25. August 1941 |
| | Document for the badge: | 5. September 1941 |
| **Wolfgang Grimm:** | Glider Pilot License (L): | 1. September 1941 |
| | Document for the badge: | 4. October 1941 |
| **Karl-Heinz Gerstmann:** | Glider Pilot License (L): | 15. September 1941 |
| | Document for the badge: | 30. September 1941 |

Each case shows a different length of the processing. Essentially the date of the glider pilot license (L) is always earlier than the one on the badge document. However, there are a few exceptions as the following example shows:

| **Erich Ullrich** | Glider Pilot License (L): | 30. June 1942 |
|---|---|---|
| | Document for the badge: | 20. August 1941 |

In this case the reason was that the glider pilot license was a duplicate (*Zweitausfertigung*) with a new number for the original but lost example. (Another such example of this circumstance was Anton Röhrle).

On each document for the glider pilot badge one can find a number in the lower left corner. For example: Ludwig Egner : 67034/41 or Karl Sauer: 83161/41

These consecutive numbers were given out from the *Reichtsluftfahrtsministerium* (RLM) beginning in the year 1935. Each member of the flying personnel who successfully passed a test - be it as a pilot, observer, air gunner, glider pilot - got a corresponding document. So the number on the document for Ludwig Egner means that he was the 67034th member of the flying personnel of the *Luftwaffe* who successfully passed a qualification test.

Only for the flyer's commemorative badge and the parachutist's badge (both *Luftwaffe* and *Heer)* was a different numbering system was used.

On the following pages some of the original documents in the format A4 and A5 will be shown. The collector and interested students of the subject can easily distinguish the differences between the document types, which are not restricted to the format alone.

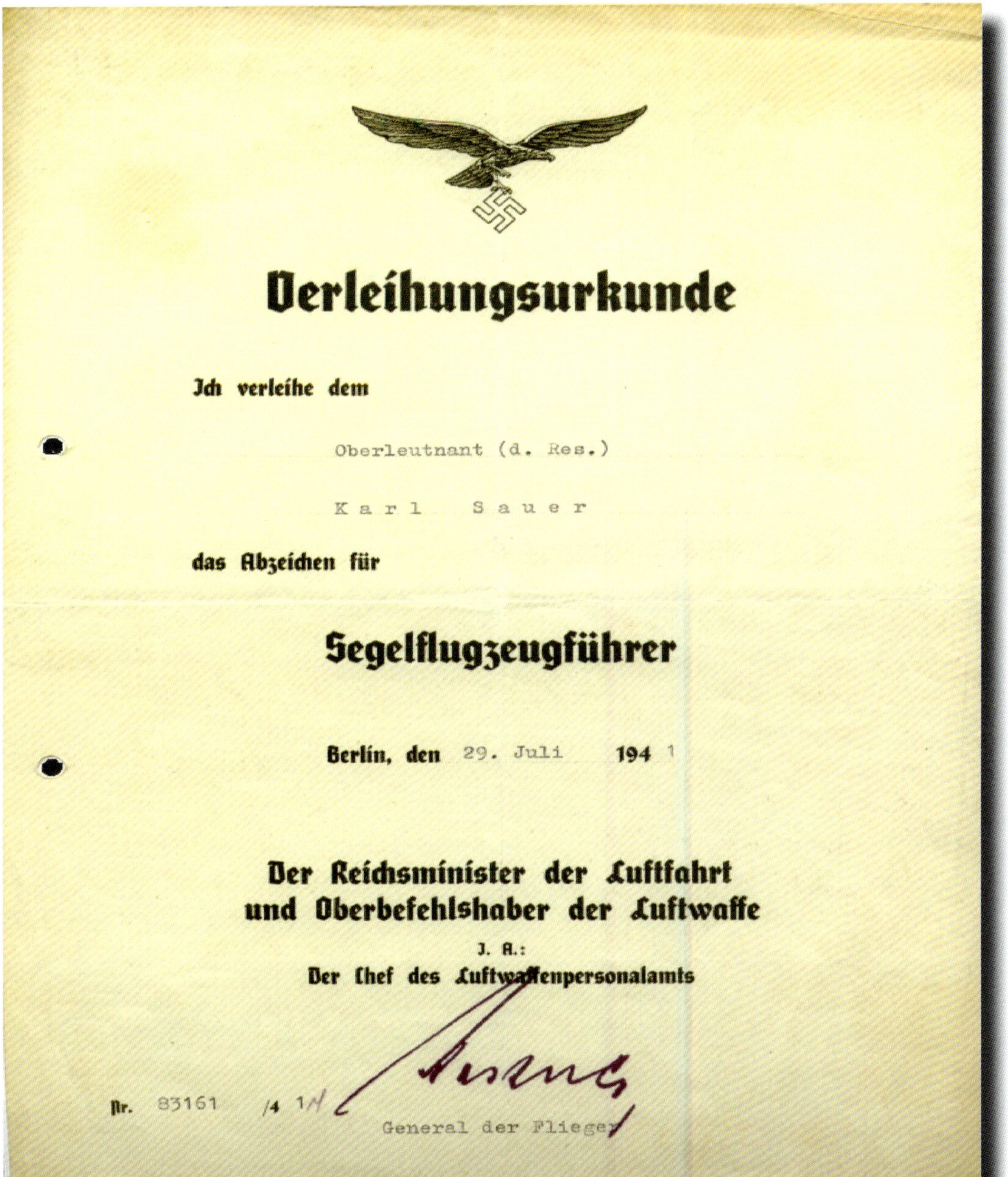

# Verleihungsurkunde

Ich verleihe dem

Oberleutnant (d. Res.)

K a r l   S a u e r

das Abzeichen für

## Segelflugzeugführer

Berlin, den 29. Juli 1941

**Der Reichsminister der Luftfahrt
und Oberbefehlshaber der Luftwaffe**

I. A.:
Der Chef des Luftwaffenpersonalamts

Nr. 83161 /41

General der Flieger

Document of the Type 1 (DIN A4), awarded on 29. July 1941 to *Oberleutnant* (d. Res.) Karl Sauer, Number 83161/41.

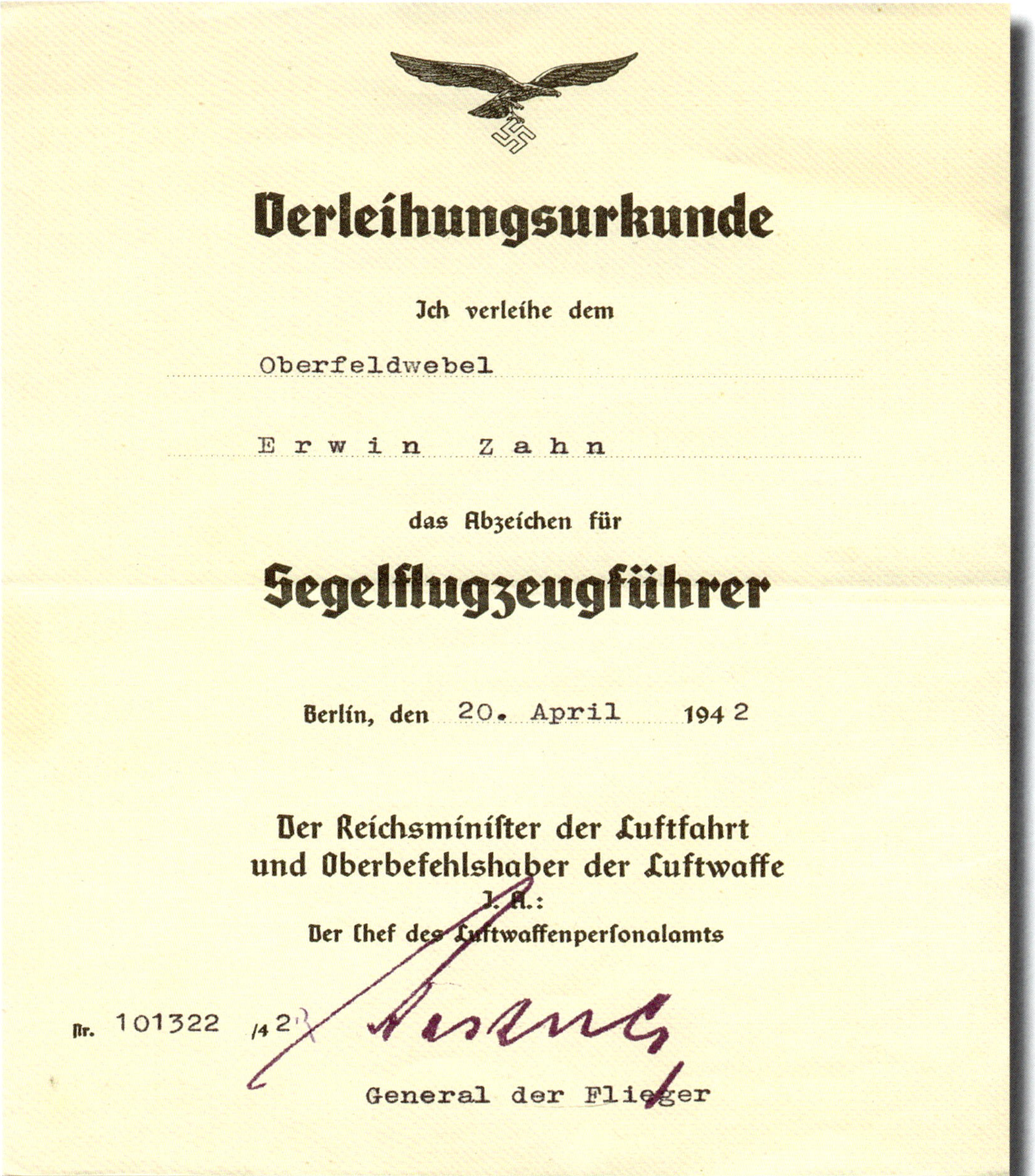

# Verleihungsurkunde

Ich verleihe dem

Oberfeldwebel

Erwin Zahn

das Abzeichen für

# Segelflugzeugführer

Berlin, den 20. April 1942

Der Reichsminister der Luftfahrt
und Oberbefehlshaber der Luftwaffe
I. A.:
Der Chef des Luftwaffenpersonalamts

Nr. 101322 /42

General der Flieger

Document of the Type 2 (DIN A5), awarded on 20. April 194 to *Oberfeldwebel* Erwin Zahn, Number 101322/42.

## A Comparison Between The Two Document Types

Not only is the format different between the two document types but also the general design. In a direct comparison these differences will be made clear. First the top part:

Verleihungsurkunde

Ich verleihe dem

Above is the top part of a A4 (Type 1) sized document. The eagle is much larger than the one on the A5 document (below) but is of the same design down to the finest detail. The other major difference is that the sentence "*Ich verleihe dem*" ("I award to") is centered in the middle of the document with Type 2 (A5)

Verleihungsurkunde

Ich verleihe dem

(Both document portions are not to scale, but the size relation to each other is correct.)

Here are the middle portions of the same documents:

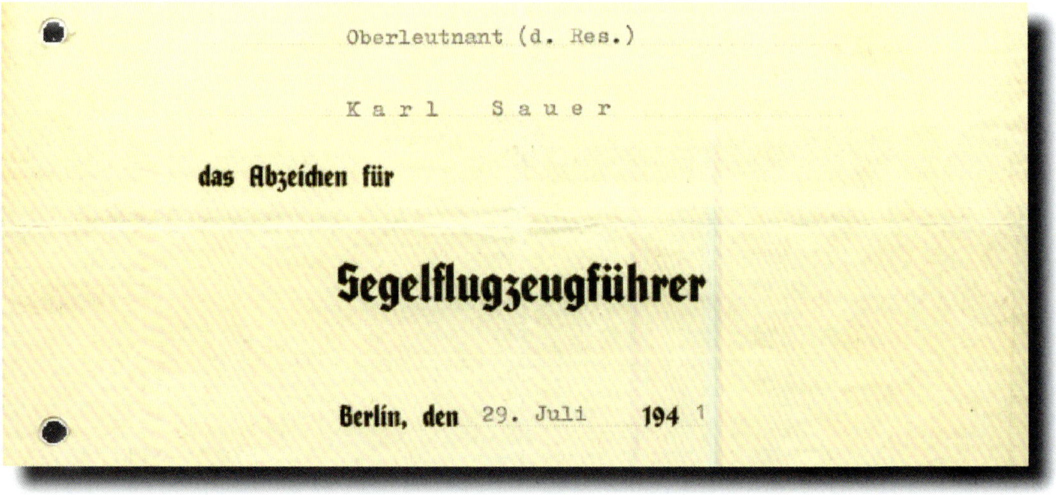
Oberleutnant (d. Res.)

Karl Sauer

das Abzeichen für

Segelflugzeugführer

Berlin, den 29. Juli 1941

Oberfeldwebel

Erwin Zahn

das Abzeichen für

Segelflugzeugführer

Berlin, den 20. April 1942

Apart from the size of the documents the main difference is again that the Type 2 (A5) has the "*das Abzeichen für*" ("the badge for") centered. Another difference is that the Type 1 (A4) has the date line "Berlin, den ............194 " in bold letters compared to the regular fonts of the Type 2.

The bottom portions of the same documents:

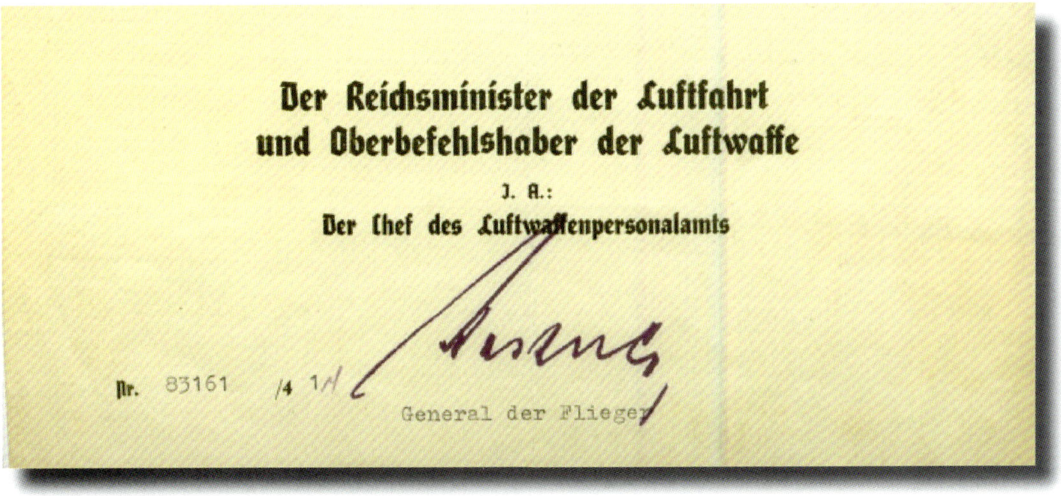

Der Reichsminister der Luftfahrt
und Oberbefehlshaber der Luftwaffe
J. A.:
Der Chef des Luftwaffenpersonalamts

Nr. 83161 /4 1/1

General der Flieger

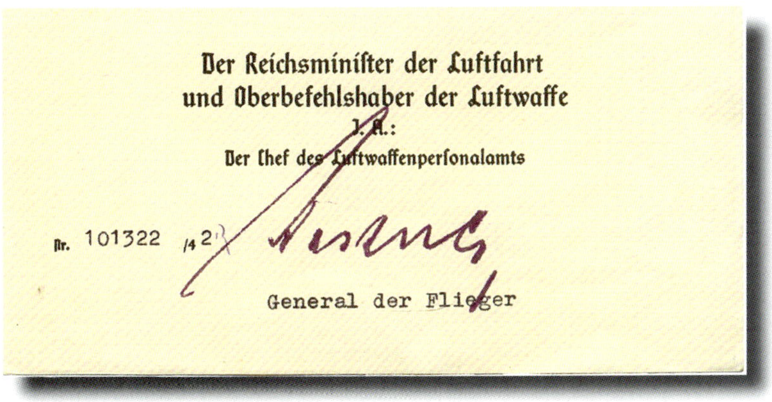

Der Reichsminister der Luftfahrt
und Oberbefehlshaber der Luftwaffe
J. A.:
Der Chef des Luftwaffenpersonalamts

Nr. 101322 /4 2

General der Flieger

There are no real obvious differences other than that the font of the Type 1 is a little more bold than that of the Type 2. Both scripts are centered.

On the following pages a few original documents of both types are shown.

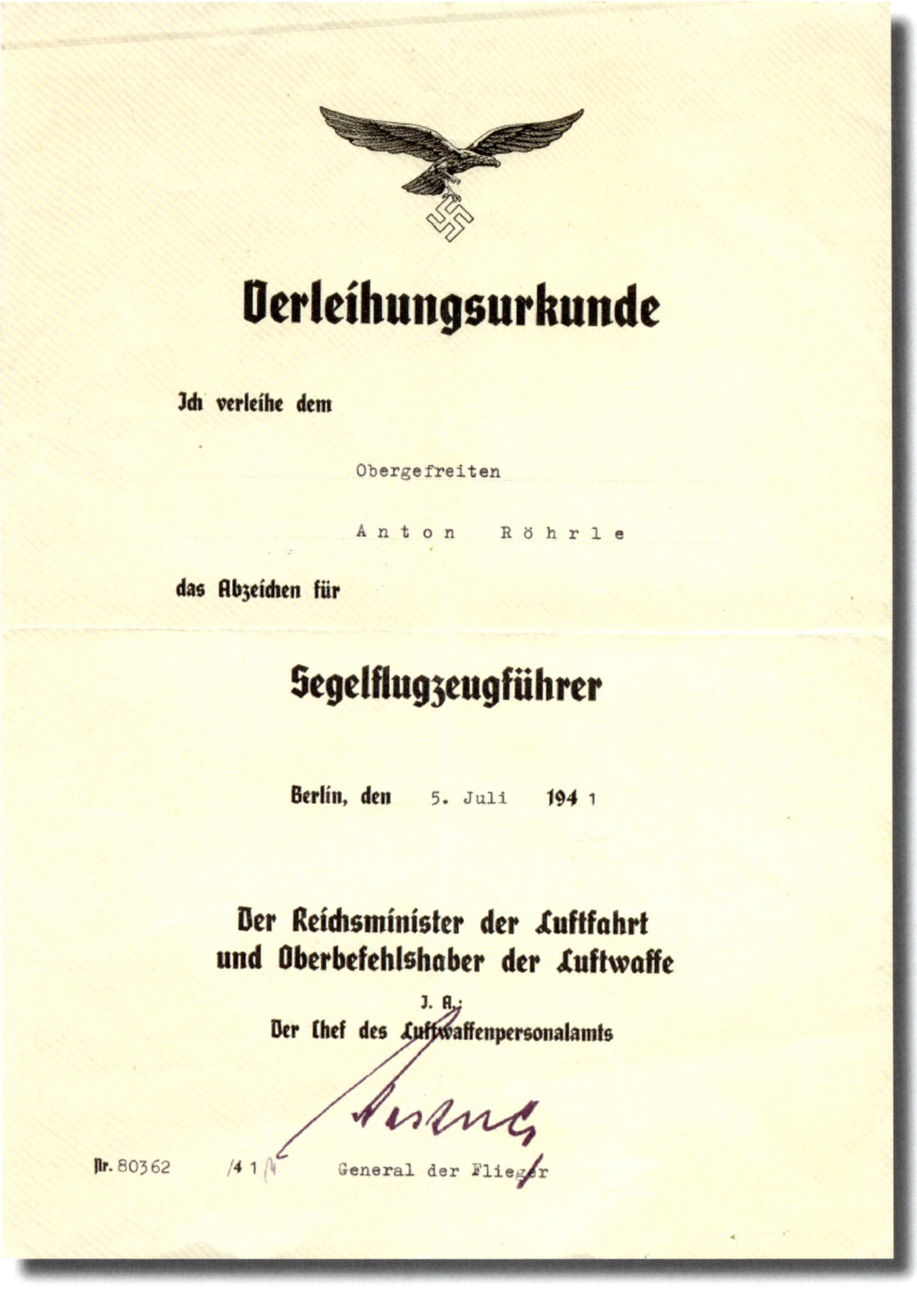

# Verleihungsurkunde

Ich verleihe dem

Obergefreiten

A n t o n   R ö h r l e

das Abzeichen für

**Segelflugzeugführer**

Berlin, den 5. Juli 1941

**Der Reichsminister der Luftfahrt und Oberbefehlshaber der Luftwaffe**

J. A.:

Der Chef des Luftwaffenpersonalamts

Nr. 80362 /41

General der Flieger

Document Type 1 (A4) awarded to *Obergefreiter* Anton Röhrle, 5 July 1941.

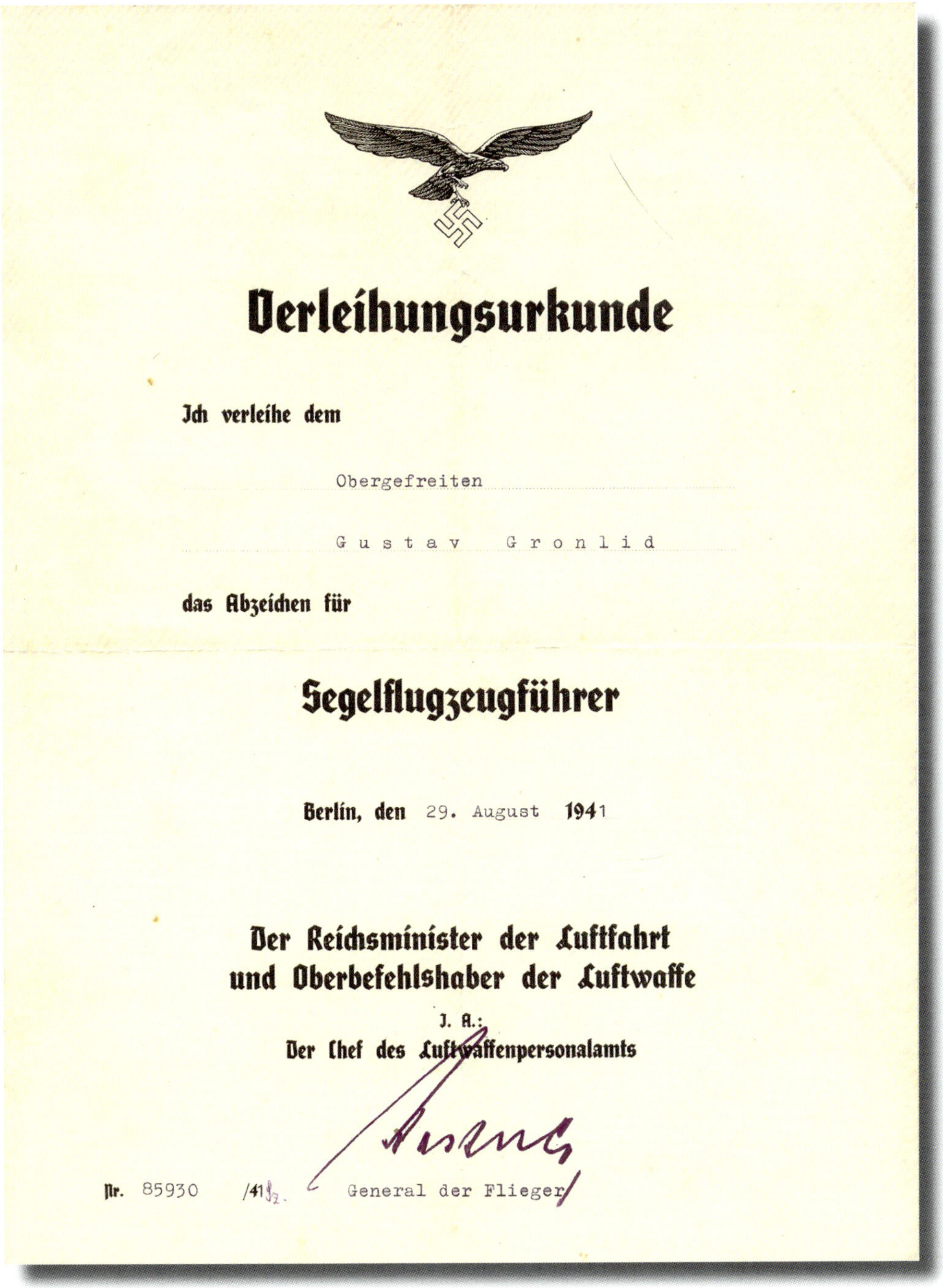

Verleihungsurkunde

Ich verleihe dem

Obergefreiten

G u s t a v   G r o n l i d

das Abzeichen für

Segelflugzeugführer

Berlin, den 29. August 1941

Der Reichsminister der Luftfahrt
und Oberbefehlshaber der Luftwaffe

I. A.:
Der Chef des Luftwaffenpersonalamts

Nr. 85930 /41 General der Flieger

Document Type 1 (A4) awarded to *Obergefreiter* Gustav Gronlid, 29. August 1941.

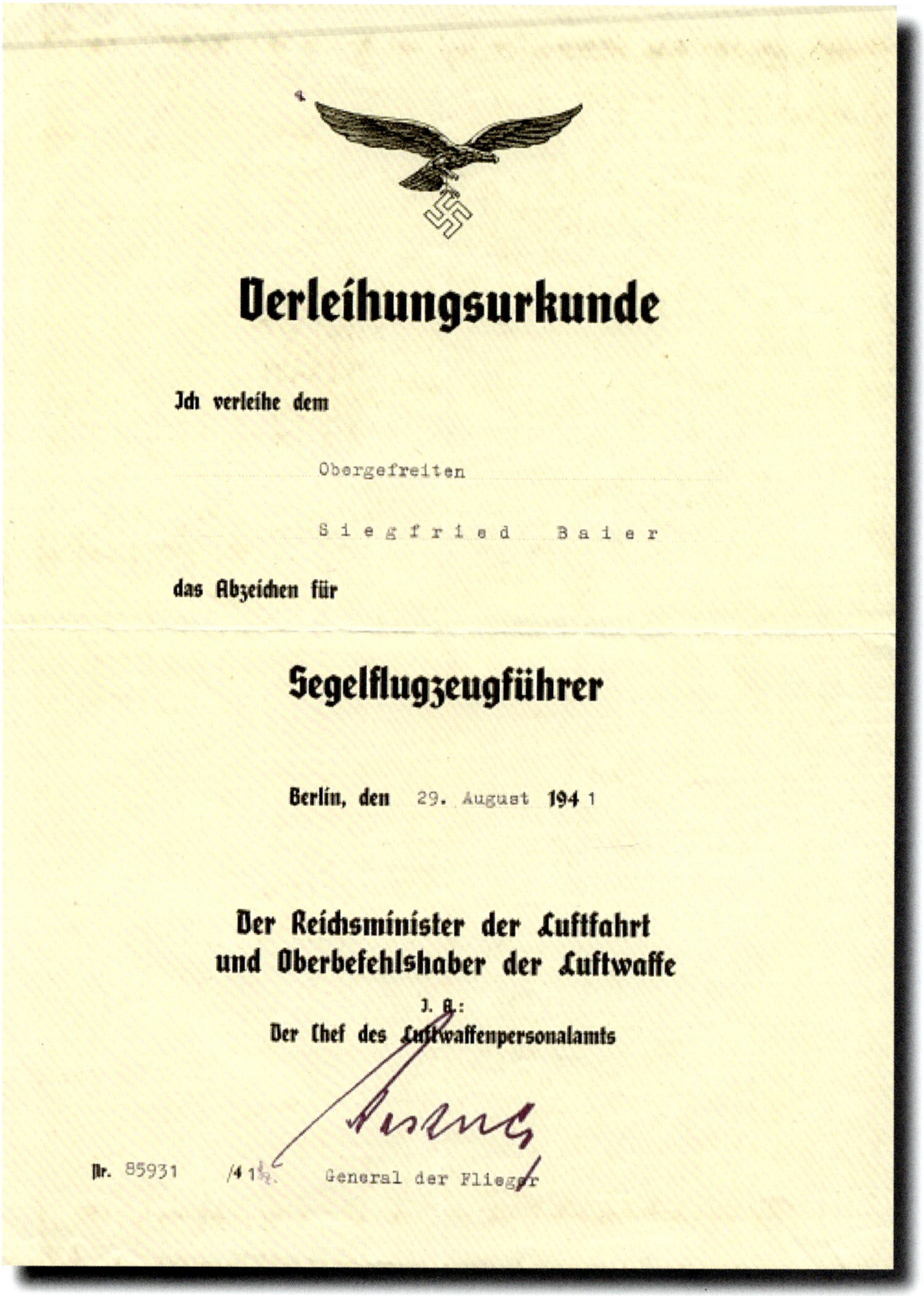

Verleihungsurkunde

Ich verleihe dem

Obergefreiten

S i e g f r i e d   B a i e r

das Abzeichen für

Segelflugzeugführer

Berlin, den 29. August 1941

Der Reichsminister der Luftfahrt
und Oberbefehlshaber der Luftwaffe

I. A.:
Der Chef des Luftwaffenpersonalamts

Nr. 85931 /41 L.   General der Flieger

Document Type 1 (A4) awarded to *Obergefreiter* Siegfried Baier, 29. August 1941.

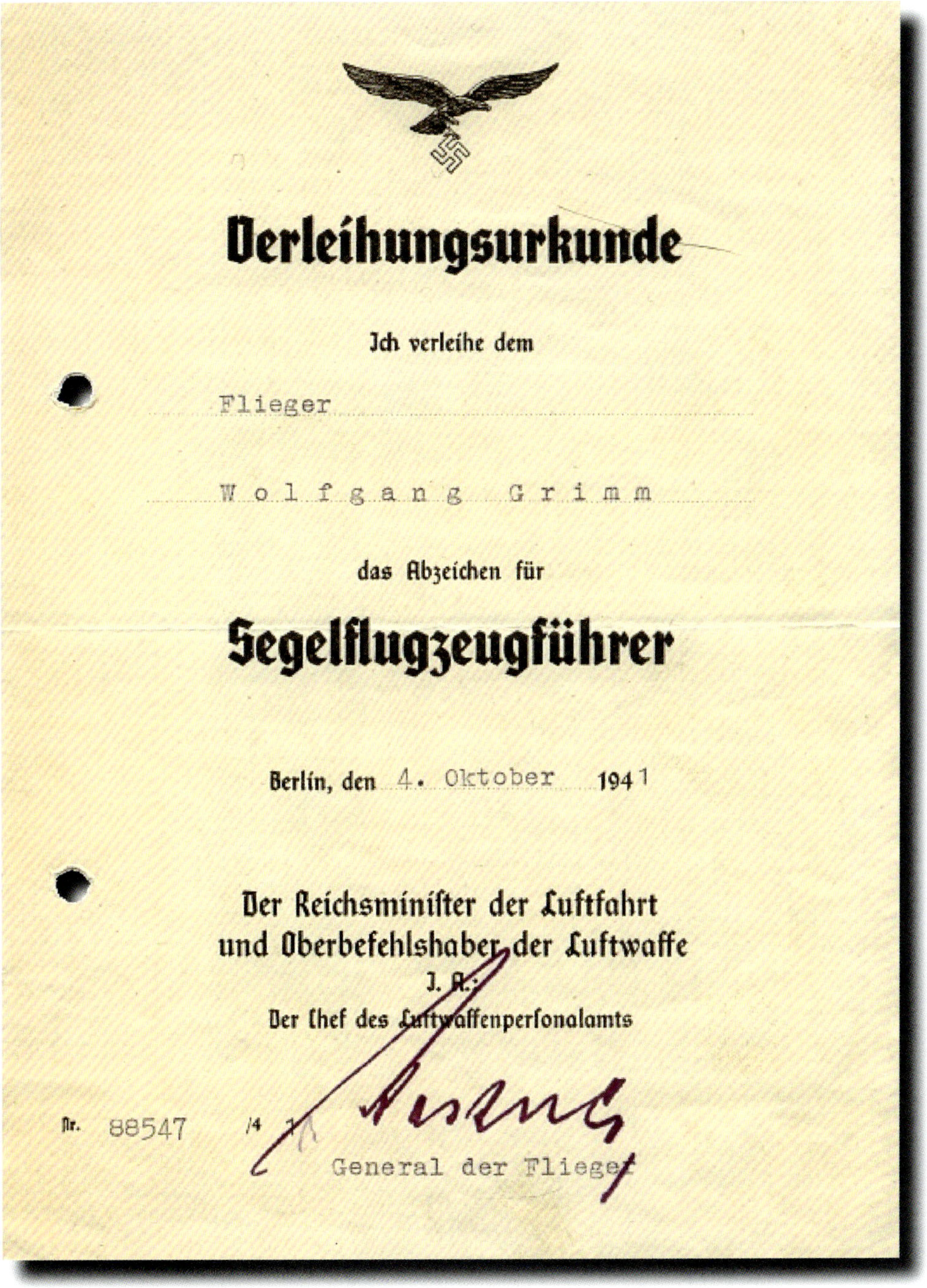

# Verleihungsurkunde

Ich verleihe dem

Flieger

W o l f g a n g   G r i m m

das Abzeichen für

# Segelflugzeugführer

Berlin, den 4. Oktober 1941

Der Reichsminister der Luftfahrt
und Oberbefehlshaber der Luftwaffe
I. A.
Der Chef des Luftwaffenpersonalamts

Nr. 88547 /41

General der Flieger

Document Type 2 (A5) awarded to *Flieger* Wolfgang Grimm, 4. October 1941.

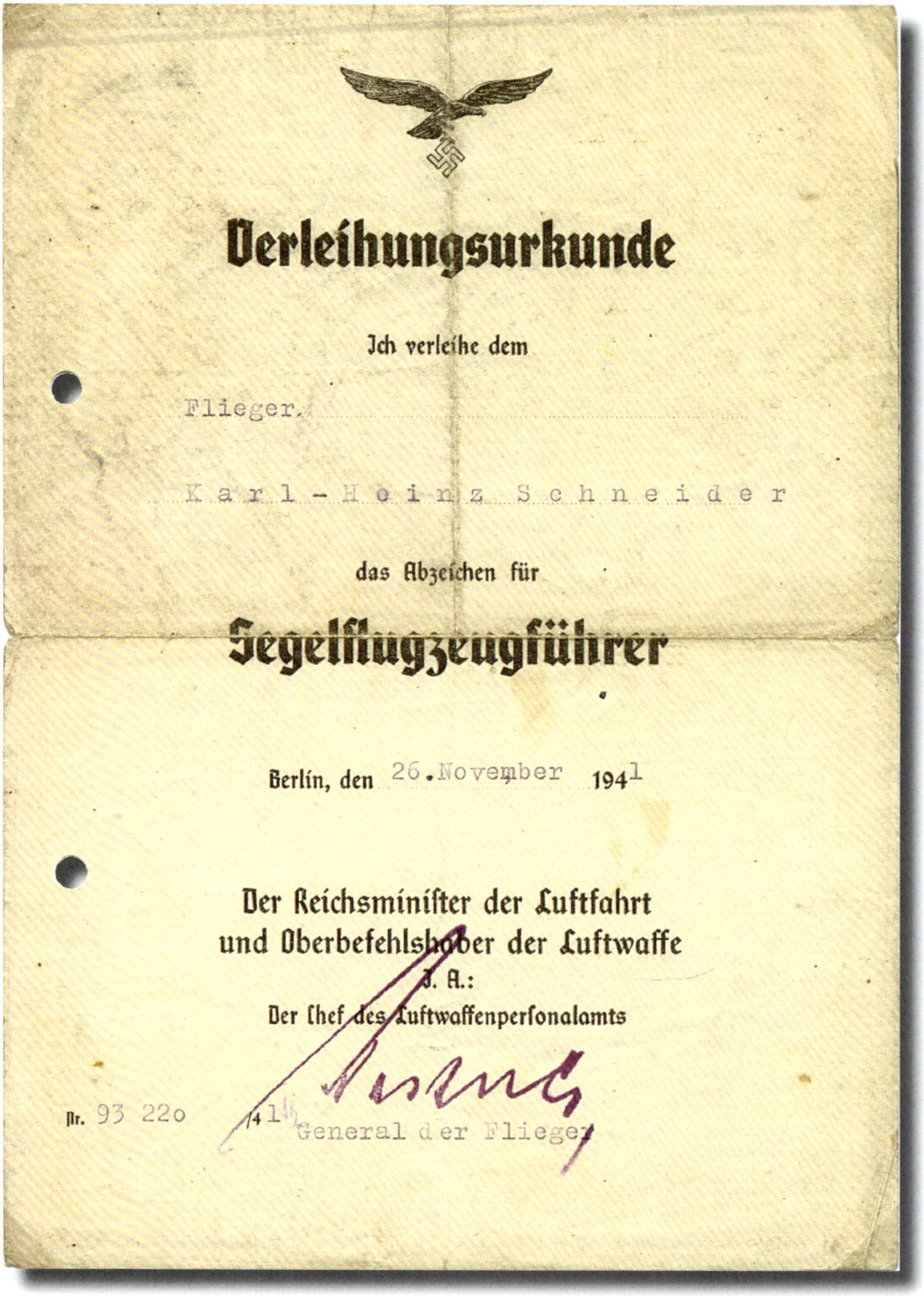

# Verleihungsurkunde

Ich verleihe dem

Flieger

Karl-Heinz Schneider

das Abzeichen für

# Segelflugzeugführer

Berlin, den 26. November 1941

Der Reichsminister der Luftfahrt
und Oberbefehlshaber der Luftwaffe
J. A.:
Der Chef des Luftwaffenpersonalamts

Nr. 93 220 /41 General der Flieger

Document Type 2 (A5) awarded to *Flieger* Karl-Heinz Schneider, 26. November 1941.

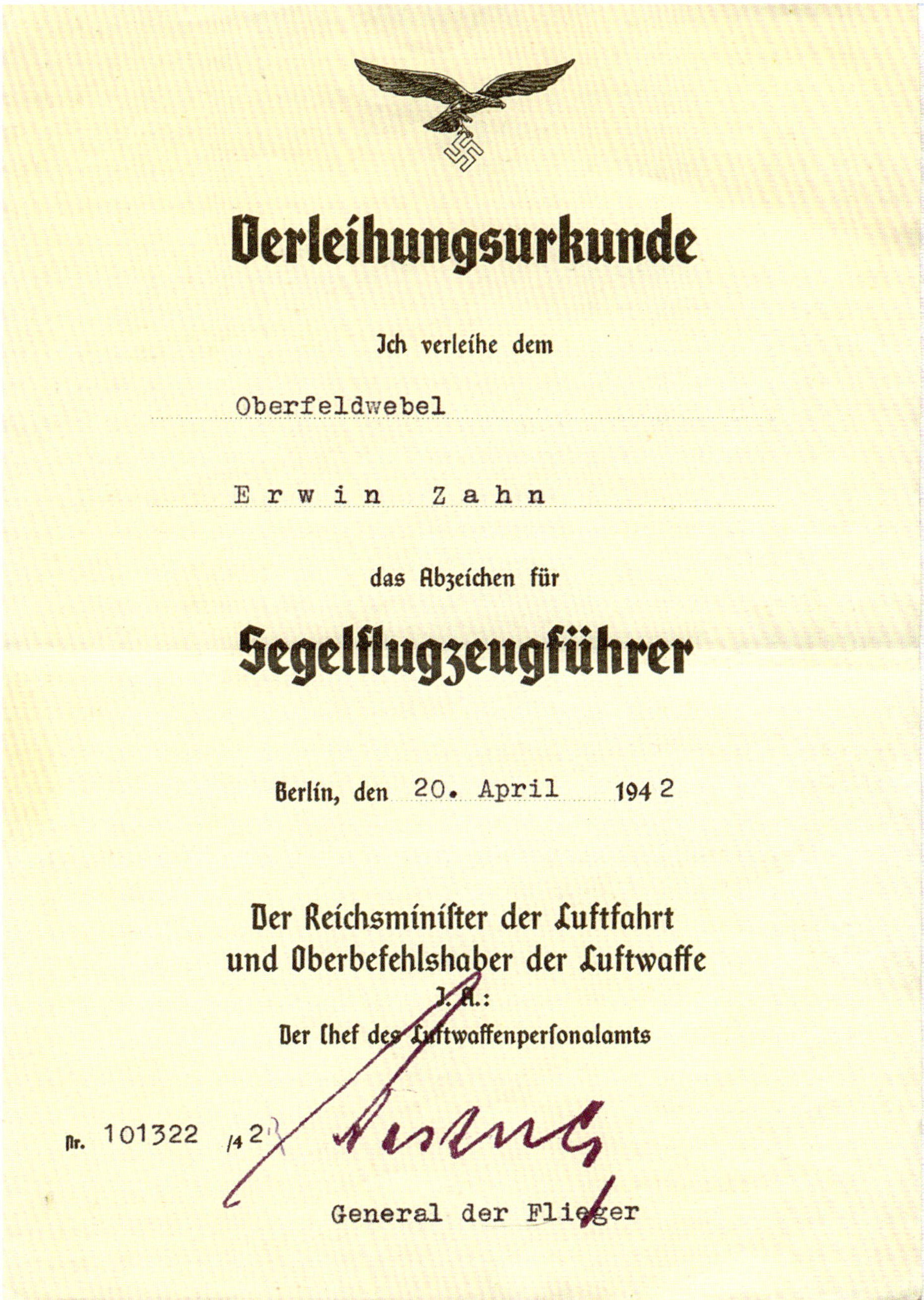

Verleihungsurkunde

Ich verleihe dem

Oberfeldwebel

Erwin Zahn

das Abzeichen für

Segelflugzeugführer

Berlin, den 20. April 1942

Der Reichsminister der Luftfahrt
und Oberbefehlshaber der Luftwaffe
I. A.:
Der Chef des Luftwaffenpersonalamts

Nr. 101322 /42

General der Flieger

Document Type 2 (A5) awarded to *Oberfeldwebel* Erwin Zahn, 20. April 1942.

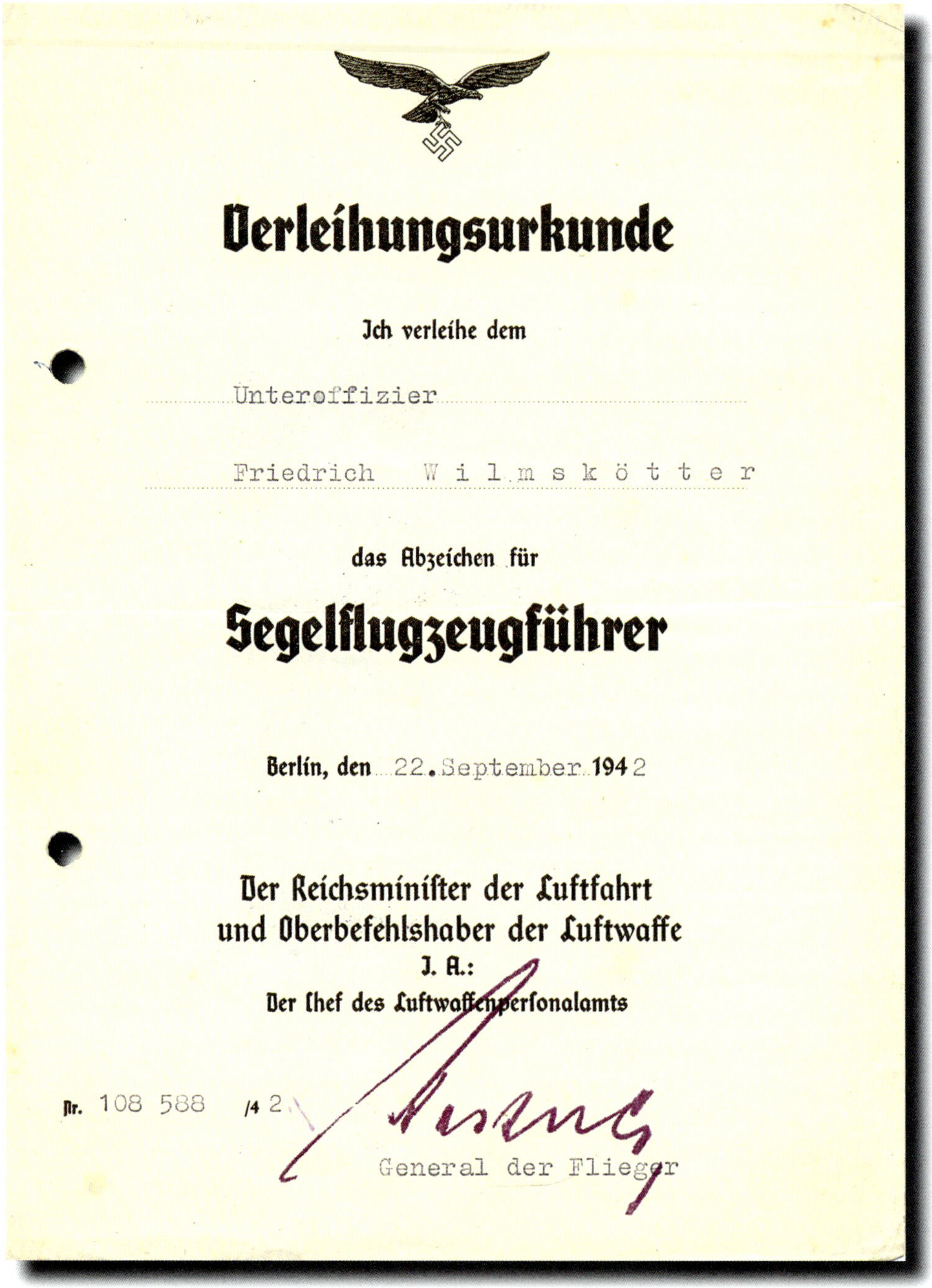

# Verleihungsurkunde

Ich verleihe dem

Unteroffizier

Friedrich W i l m s k ö t t e r

das Abzeichen für

# Segelflugzeugführer

Berlin, den 22. September 1942

Der Reichsminister der Luftfahrt
und Oberbefehlshaber der Luftwaffe
I. A.:
Der Chef des Luftwaffenpersonalamts

Nr. 108 588 /42

General der Flieger

Document Type 2 (A5) awarded to *Unteroffizier* Friedrich Wilmskötter, 22. Sept. 1942.

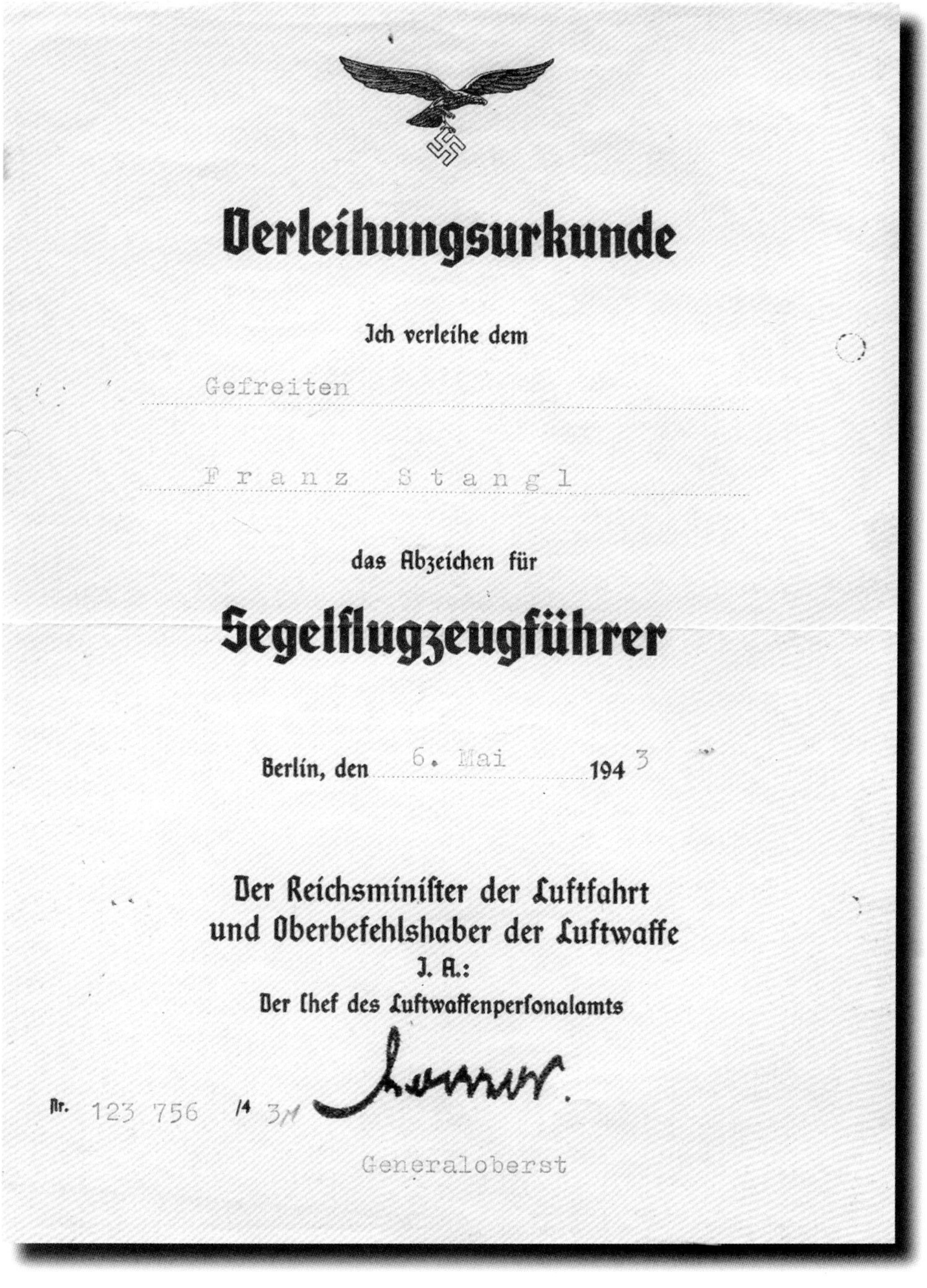

**Verleihungsurkunde**

Ich verleihe dem

Gefreiten

Franz Stangl

das Abzeichen für

**Segelflugzeugführer**

Berlin, den 6. Mai 1943

Der Reichsminister der Luftfahrt
und Oberbefehlshaber der Luftwaffe
I. A.:
Der Chef des Luftwaffenpersonalamts

Nr. 123 756 /43

Generaloberst

Document Type 2 (A5) awarded to *Gefreiter* Franz Stangl, 6. Mai 1943.

# THE LUFTWAFFE GLIDER PILOT BADGE IN THE PERSONNEL PAPERS

As was dictated by the military regulations, the award of the Glider Pilot Badge was also recorded in the military identification papers of the soldier.

## The *Wehrpass* (Service Record)

The award of a *Luftwaffe* Glider Pilot Badge was always recorded in the *Wehrpass* on page 21.

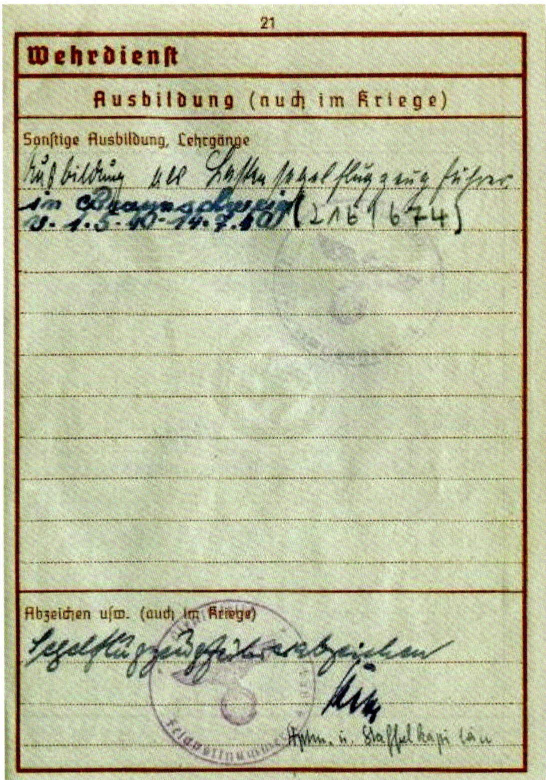

21

Wehrdienst

Ausbildung (auch im Kriege)

Sonstige Ausbildung, Lehrgänge

(216/674)

Abzeichen usw. (auch im Kriege)

Segelflugzeugführerabzeichen

The picture (left) shows page 21 of the *Wehrpass* of the *Unteroffizier* Heinrich Neuhaus. He was drafted into the *Luftwaffe* on 1. May 1940 and immediately trained as a glider pilot and was therefore one of the earliest.

After the successful completion of his training he was awarded the *Luftwaffe* Glider Pilot Badge on 29. May 1941.

The handwritten entry *Lastensegelflugzeugführer* (216/674) is already an indication that he was trained to become a military glider pilot. Most of the service records show this entry in the form of a stamp (refer to the *Wehrpass* of Karl-Heinz Gerstmann).

The picture (below) shows the detail of the page (left) with the entry of the "*Segelflugzeugführerabzeichen*" (Glider Pilot Badge).

Abzeichen usw. (auch im Kriege)

Segelflugzeugführerabzeichen

The picture (right) again shows the page 21 of a *Wehrpass*, this time that of *Leutnant* H. Becher who was eligible to wear the Glider Pilot Badge.

The entry in his record is atypical since it says "*Segelflugzeugführer (L)*" instead of the usual entry "*Flugzeugführer (L)* (674)."

Further confusion might arise due to the misleading entry "*Flugzeugführer: Abzeichen* (L)". This soldier clearly was a pilot of a cargo glider, and this entry could most likely be traced back to an office clerk with wanting knowledge who wrote this entry.

Another such example can be found in the *Wehrpass* of *Oberfeldwebel* Erwin Zahn.

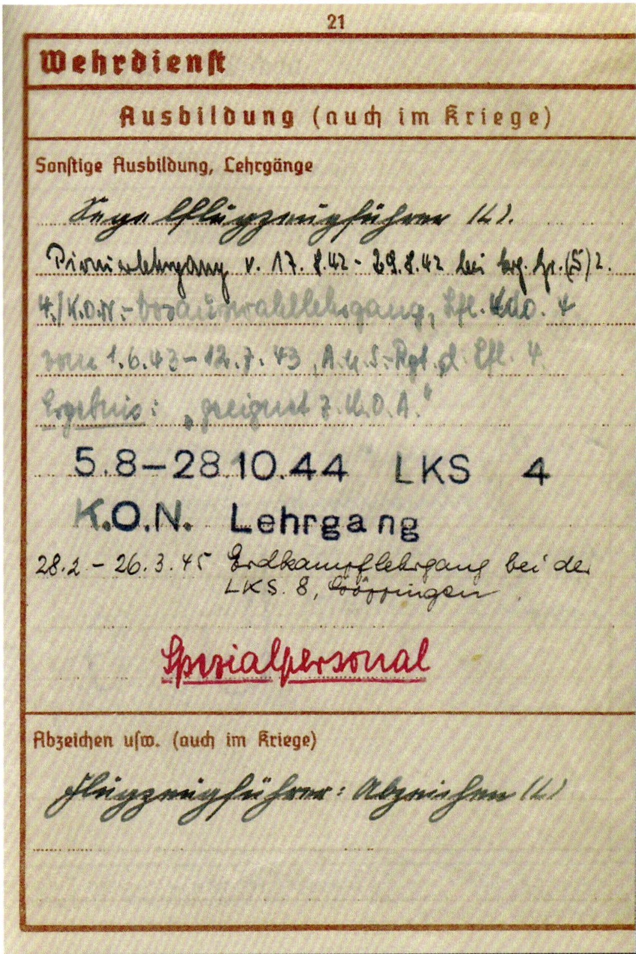

21

Wehrdienst

Ausbildung (auch im Kriege)

Sonstige Ausbildung, Lehrgänge

5.8–28.10.44 LKS 4

K.O.N. Lehrgang

Abzeichen usw. (auch im Kriege)

The detail picture (below) of the entry "*Flugzeugführer: Abzeichen (L).*"

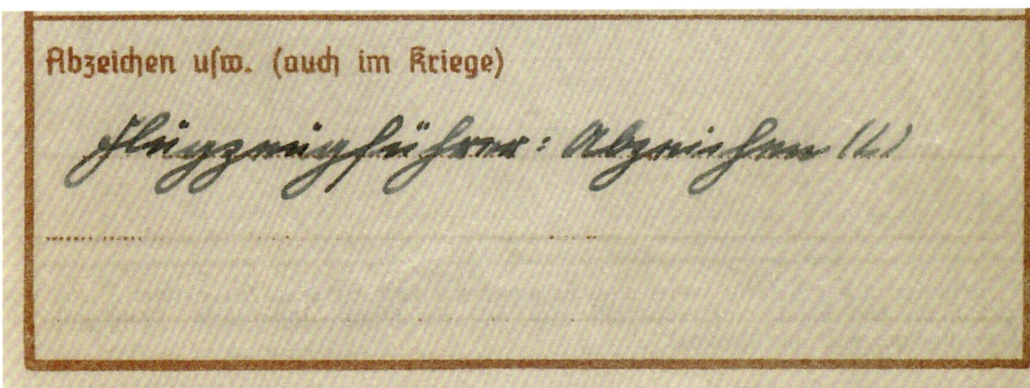

Abzeichen usw. (auch im Kriege)

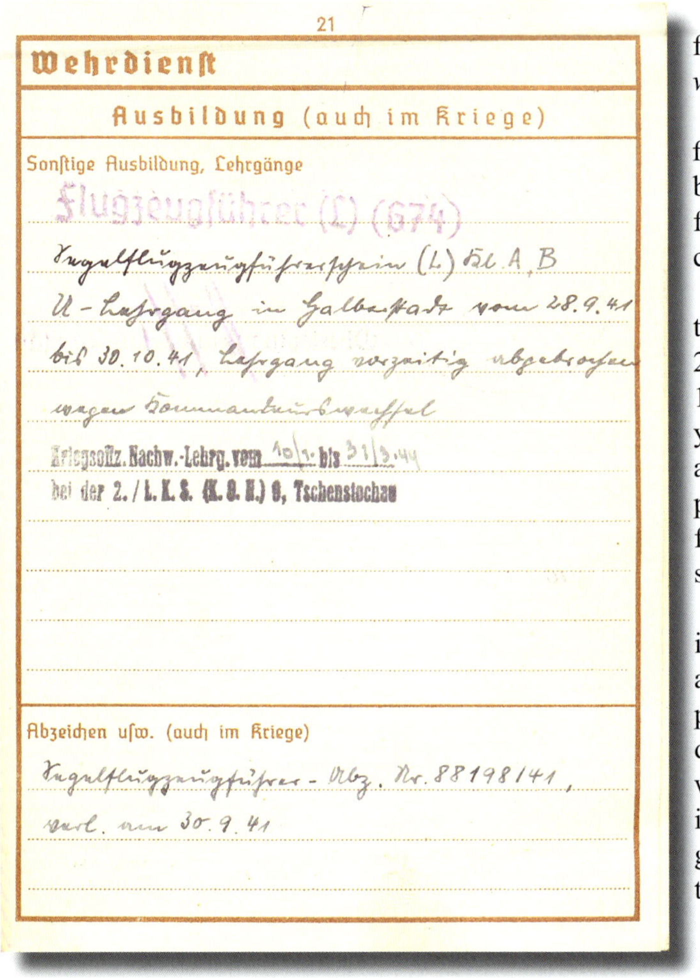
21

**Wehrdienst**

Ausbildung (auch im Kriege)

Sonstige Ausbildung, Lehrgänge

Flugzeugführer (L) (674)

Segelflugzeugführerschein (L) Kl. A, B

U-Lehrgang in Halberstadt vom 28.9.41

bis 30.10.41, Lehrgang vorzeitig abgebrochen

wegen Kommandantenwechsel

Kriegsoffz. Nachw.-Lehrg. vom 10/1. bis 31/3.44

bei der 2./L.K.S. (K.O.N.) 6, Tschenstochau

Abzeichen usw. (auch im Kriege)

Segelflugzeugführer-Abz. Nr. 88198141,

verl. am 30.9.41

The same page of the pass from *Fallschirmjäger Feldwebel* Karl-Heinz Gerstmann. It not only shows the qualification as a glider pilot (L), but there are also the entries for the glider pilot license (L) classes A and B.

Karl-Heinz Gerstmann attended an NCO course from 28. September to 30. October 1941 in Halberstadt. In the year 1944 he was assigned to a war officer course, but the promotion did not come before war's end when he was still a *Feldwebel.*

K.-H. Gerstmann was active in the Mediterranean theater and it was there that he completed 45 combat flights under difficult circumstances. He was awarded the flying clasp in bronze for transport and glider squadrons on 1. September 1942.

Detail from page 21 showing the entry for the pilot glider badge and the interesting reference to the document number and the document date.

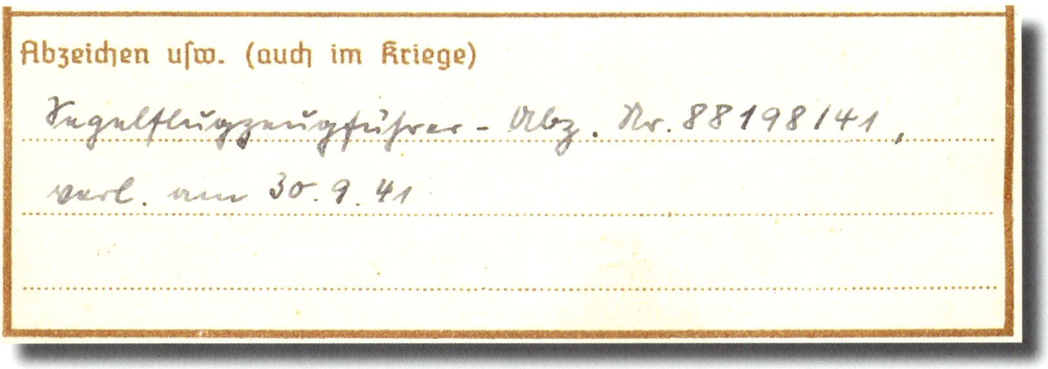
Abzeichen usw. (auch im Kriege)

Segelflugzeugführer-Abz. Nr. 88198141,

verl. am 30.9.41

The Pass of *Leutnant* Grimm shows the typical determination *Flugzeugführer (L) (674)* in the first line. Below is an entry for the participation of a war officer course from 10. January to 31. March 1944 in the town of Tschenstochau/ Silesia.

It is very interesting to find as a last entry the remark "*Flugzeugführer Nr.* 1210/45." *Leutnant* Grimm was mainly deployed on the eastern front. At the end of the war he trained as a regular (motor driven) airplane pilot.

He survived the war relatively unharmed.

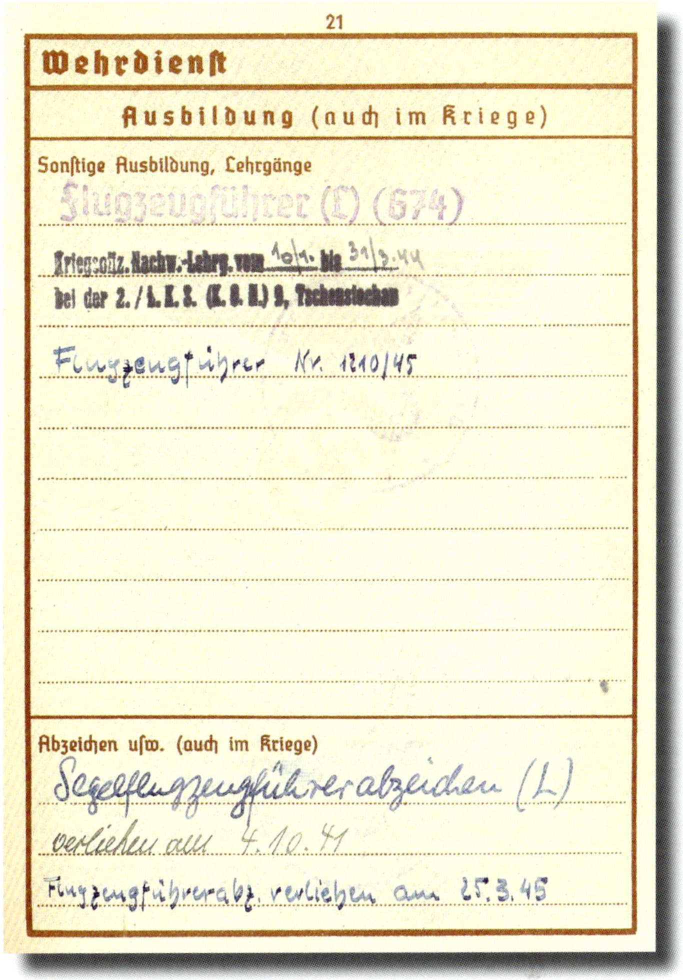

21

**Wehrdienst**

**Ausbildung (auch im Kriege)**

Sonstige Ausbildung, Lehrgänge

Flugzeugführer (L) (674)

Krieg.offz. Nachw.-Lehrg. vom 10/1. bis 31/3.44

bei der 2./L.K.S. (K.O.N.) 9, Tschenstochau

Flugzeugführer Nr. 1210/45

Abzeichen usw. (auch im Kriege)

Segelflugzeugführerabzeichen (L)

verliehen am 4.10.41

Flugzeugführerabz. verliehen am 25.3.45

Detail of the entries.

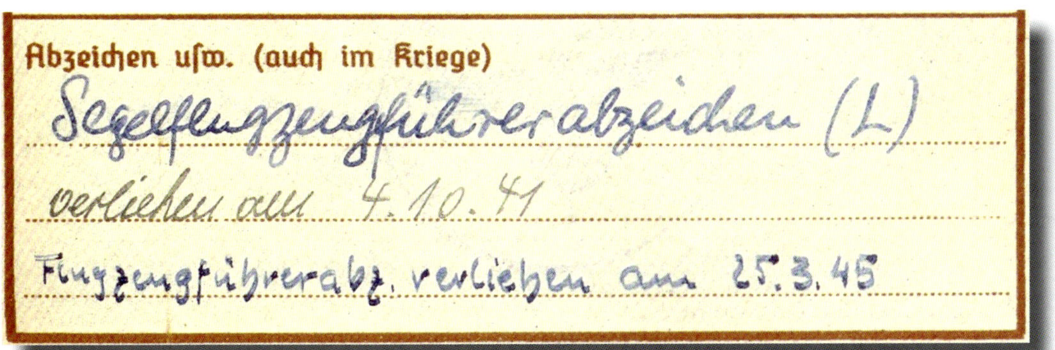

Abzeichen usw. (auch im Kriege)

Segelflugzeugführerabzeichen (L)

verliehen am 4.10.41

Flugzeugführerabz. verliehen am 25.3.45

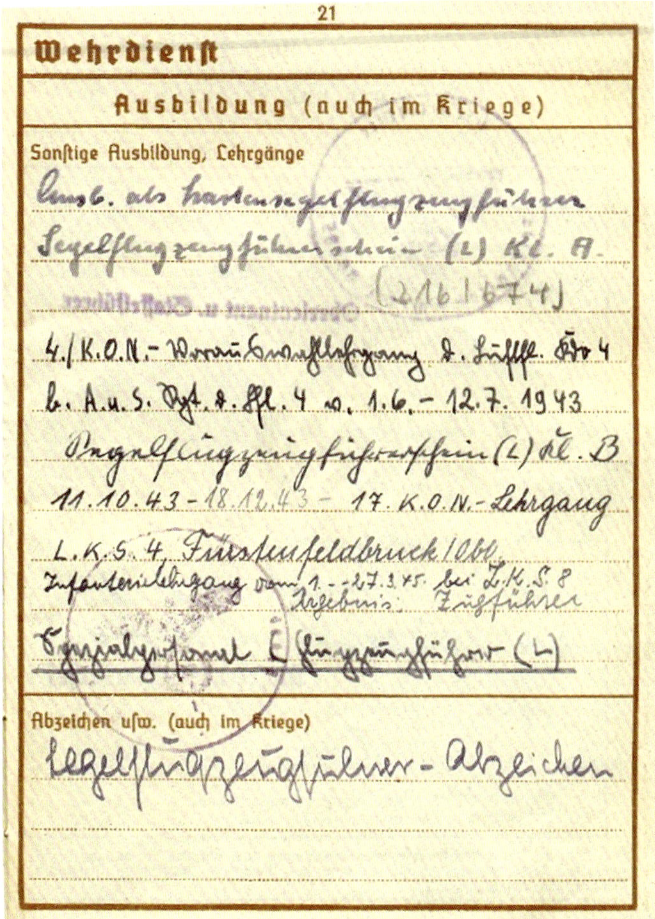

21

**Wehrdienst**

Ausbildung (auch im Kriege)

Sonstige Ausbildung, Lehrgänge

(2161674)

11.10.43 - 18.12.43 - 17. K.O.N.-Lehrgang

L.K.S. 4 Fürstenfeldbruck/Obb.

Abzeichen usw. (auch im Kriege)

Segelflugzeugführer-Abzeichen

This *Wehrpass* is interesting because of the multiple records for training.

He trained as a *Lastensegelflugzeugführer* (para glider pilot) and then *Segelflugzeugführerschein* (glider pilot licence). Both entries are in handwriting.

K.H. Schneider received his Glider Pilot Badge on 26. November 1941 (Nr. 93220/41)

After his training as a para glider pilot *Leutnant* Schneider served on the *Krim* peninsula and also in other parts of Russia. At the end of the war he was wounded and ended the war in a hospital.

The pilot glider badge was the only badge *Leutnant* Schneider got during the war.

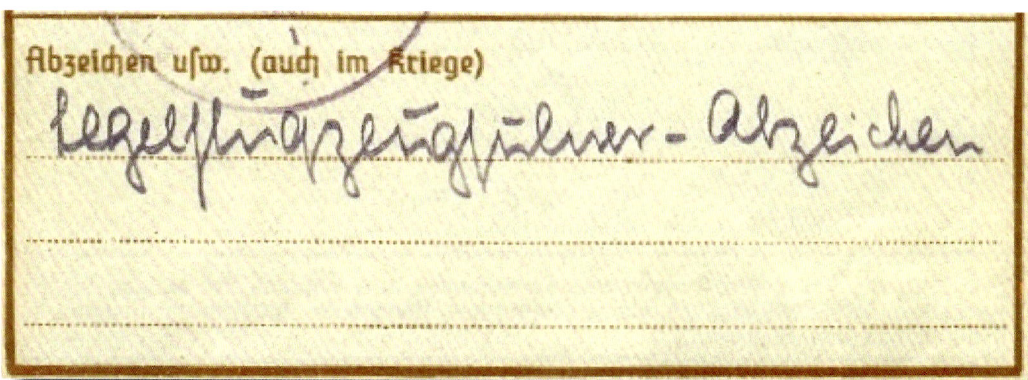

Abzeichen usw. (auch im Kriege)

Segelflugzeugführer-Abzeichen

*Stabsfeldwebel* Stumbaum joined the police at the end of 1931, and asked in the beginning of 1939 to be transferred to the *Luftwaffe*. He started his training in August 1940 at the glider school in Waggum. His *Wehrpass* shows the following manual entries:

- *Flugzeugführer (L)*
- *Segelflugzeugführer (S) 674*

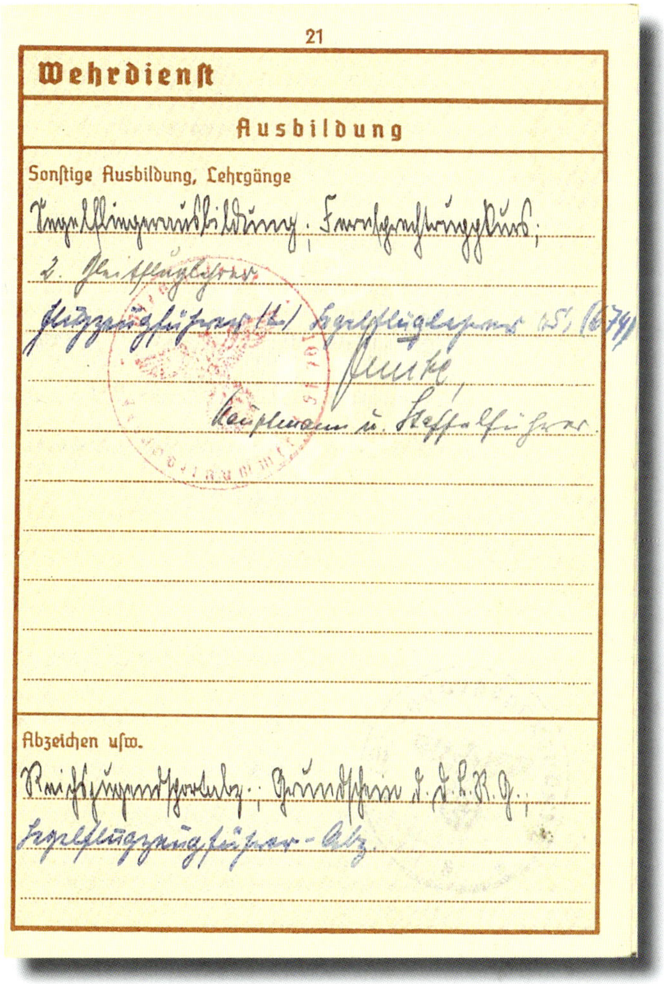
21

Wehrdienst

Ausbildung

Sonstige Ausbildung, Lehrgänge

Abzeichen usw.

During the war *Stabsfeldwebel* Stumbaum served at the Erg.Gr.(S) 1 in Langendiebach. At the end of March 1943 he was transferred to a front unit, the III/LLG 1 -*1.Staffel*.

This unit was transferred in the spring of 1943 to Croatia. *Feldwebel* Stumbaum lost his life during a glider crash on 10. May 1943.

The entry lists the *Reichsjugendsportabzeichen,* (Youth Sports badge) the *DLRG* badge (German Life Savings Organization) and the Glider Pilot Badges.

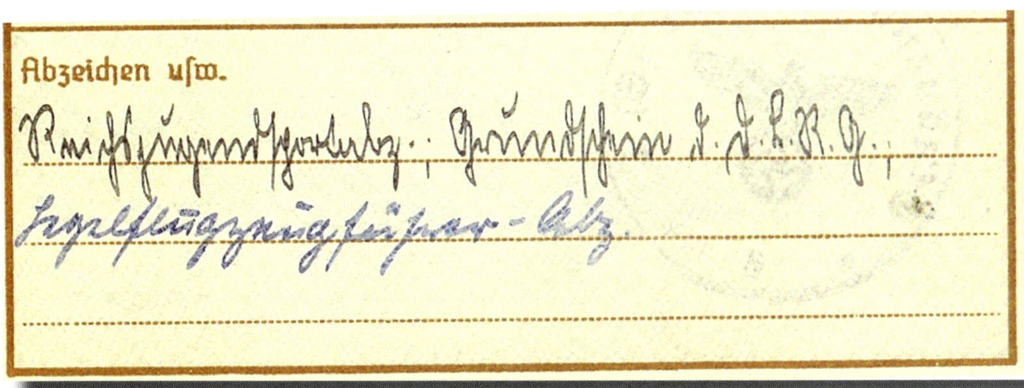
Abzeichen usw.

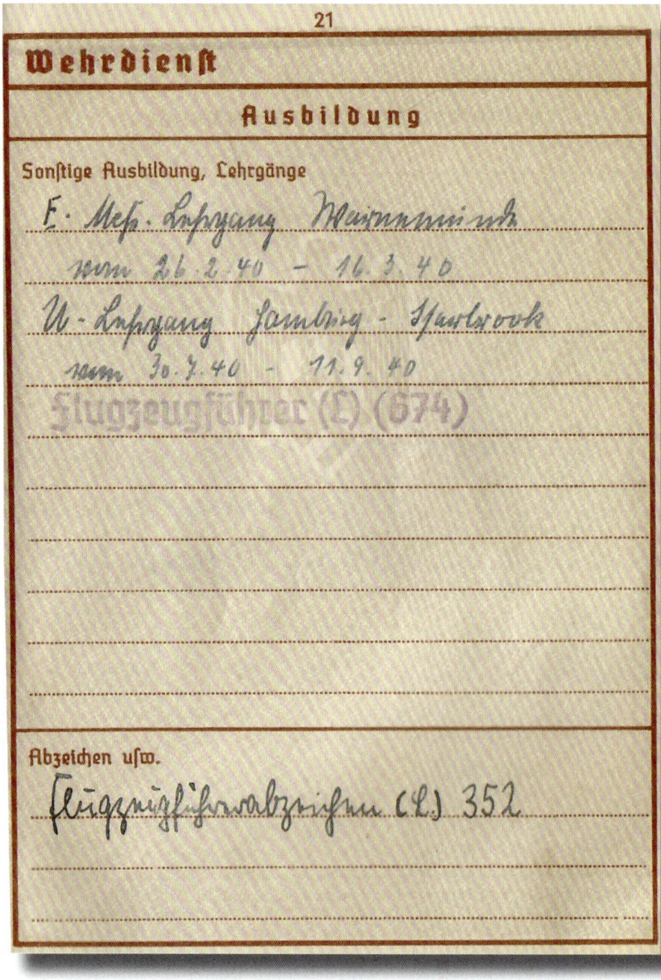
21

Wehrdienst

Ausbildung

Sonstige Ausbildung, Lehrgänge

vom 26.2.40 – 16.3.40

vom 30.7.40 – 11.9.40

Flugzeugführer (L) (674)

Abzeichen usw.

Flugzeugführerabzeichen (L) 352

The *Wehrpass* of *Oberfeldwebel* Alfred Kallmeyer also shows the typical stamped entry *Flugzeugführer (L) (674).*

However, it is very interesting to note that in the section *"Abzeichen"* (Badges) one can find the entry *Flugzeugführerabzeichen (L) 352.*

This latter entry is for the qualifying badge for piloting an engine-driven plane (i.e., it means *Motor-Flugzeugführer* instead of *Segelflugzeugführer*), not for a glider. It is not clear at this point in time whether *Ofw*. Kallmeyer trained after finishing his successful training as a glider pilot to later on become a regular pilot. (It was actually not unusual for a glider pilot to do so). *Oberfeldwebel* Kallmeyer belonged during the war to several squadrons of the LLG 1. In November 1944 he was transferred to a reconnaissance squadron where he remained until the war ended.

Kallmeyer remained an enthusiastic flyer which led to his fate. He and his son died in an airplane crash 23. September 1972.

Abzeichen usw.

Flugzeugführerabzeichen (L) 352

As is the case with the previous document, this one from *Unteroffizier* Egon Lawaetz also shows the typical entry *Flugzeugführer (L) (674).*

*Uffz.* Egon Lawaetz started his training on 20. February 1941 with the *Erg.Gr.(S) 1* and graduated successfully on 16. September 1941. Thereafter he was transferred to the unit 1 DFS/LLG1.

Egon Lawaetz was seriously wounded on 17. August 1943 during a devastating attack on the airport at Istres/Southern France. He died one day later.

He also was awarded the following medals:

- Life Saving Medal
  (11. Nov. 1940)
- Pilot Glider Badge
  (date unknown)
- Glider Clasp Bronze
  (1. May 1943)

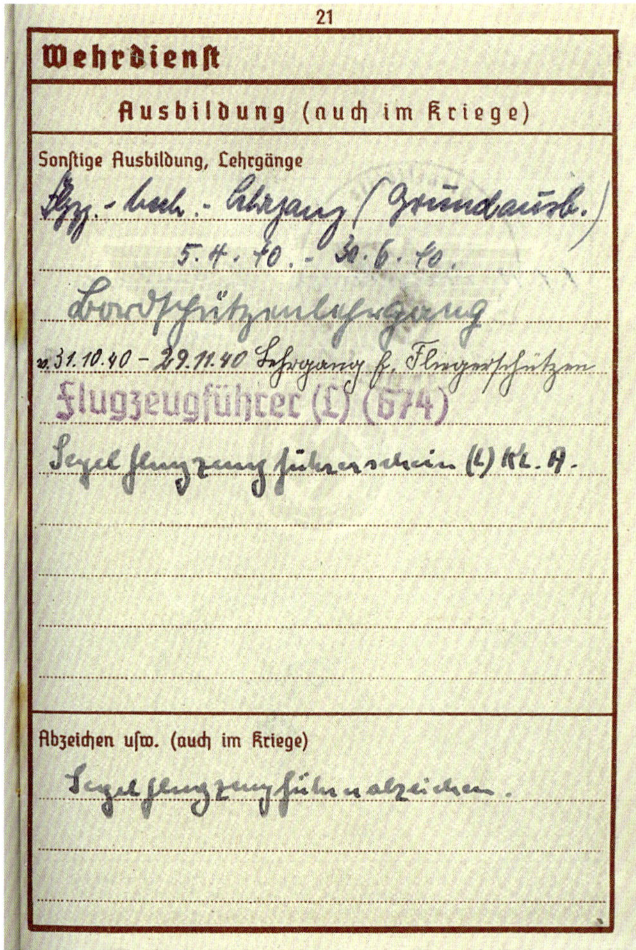
21

Wehrdienst

Ausbildung (auch im Kriege)

Sonstige Ausbildung, Lehrgänge

Flgz.-Mech.-Lehrgang (Grundausb.)
5.4.40.-30.6.40.
Bordschützenlehrgang
v. 31.10.40 - 29.11.40 Lehrgang f. Flugzeugschützen
Flugzeugführer (L) (674)
Segelflugzeugführerschein (L) Kl. A.

Abzeichen usw. (auch im Kriege)

Segelflugzeugführerabzeichen.

Abzeichen usw. (auch im Kriege)

Segelflugzeugführerabzeichen.

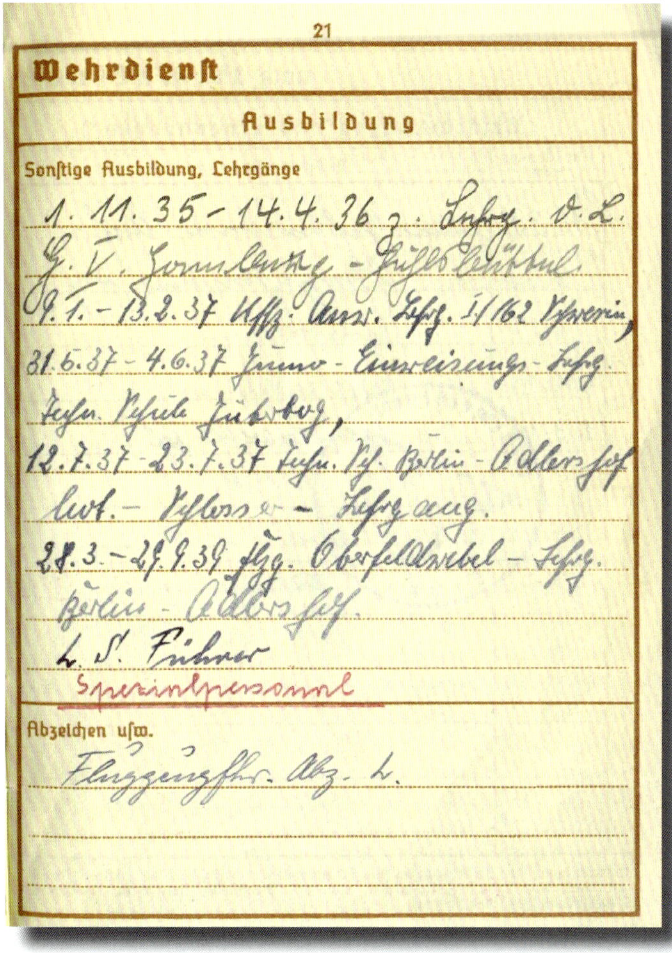

21

**Wehrdienst**

**Ausbildung**

Sonstige Ausbildung, Lehrgänge

L.S. Führer

Abzeichen usw.

Flugzeugfhr. Abz. L.

*Oberfeldwebel* Erwin Zahn was already a professional soldier during the 1930s, as one can verify with the many training courses he had.

Important for this book are only the entries:

- *L.S. Führer (***L***asten* **S***egler Führer* - para glider pilot*)*

and the slightly confusing one:

*Flugzeugfhr. Abz. L*

which is a short form of *Flugzeugführer-Abzeichen.* The "L" comes from the (L) denomination of the glider pilot license (L). One can see similar entries in the *Wehrpässe* of *Ofw.* Kallmeyer (page 70) and pay book entry of *Leutnant* K.H. Schneider.

Zahn's award document is shown on page 61. Erwin Zahn was active on the eastern front between 5. May 1943 and 3. November 1943. He and his mechanic, *Feldwebel* Hans Brandenburger, died in a crash on the later date at the airport Warwaronka/Ukraine.

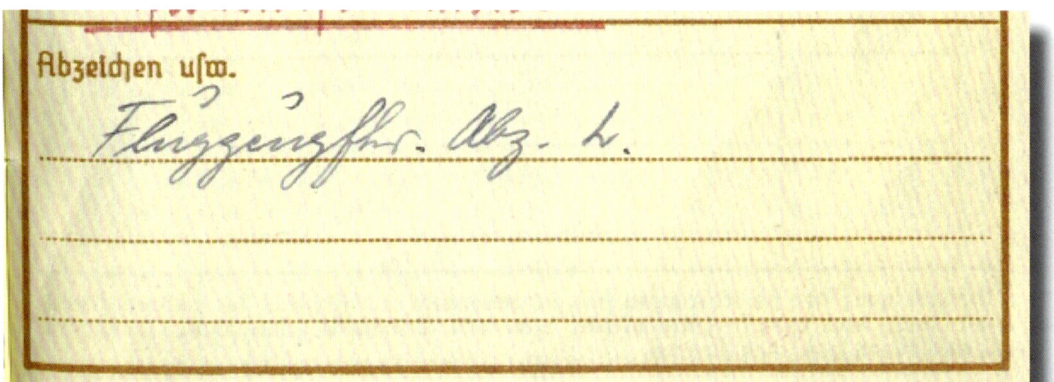

Abzeichen usw.

Flugzeugfhr. Abz. L.

*Unteroffizier-Fahnenjunker* Franz Stangl started fairly late with his glider training (March 1942) at the Erg.Gr.(S)-2 in Posen. In May of the following year he finished the training and he received his *Segelfliegerschein (L)* No. 84/43.

The document for the pilot glider badge is dated 6. May 1943 (pictures on page 63).

Remarkable is the following handwritten entry:

- *Flugzeugführer (L)*

However, the description of the glider pilot badge is correct: *Lw.-Segelflugzeugführerabzeichen.*

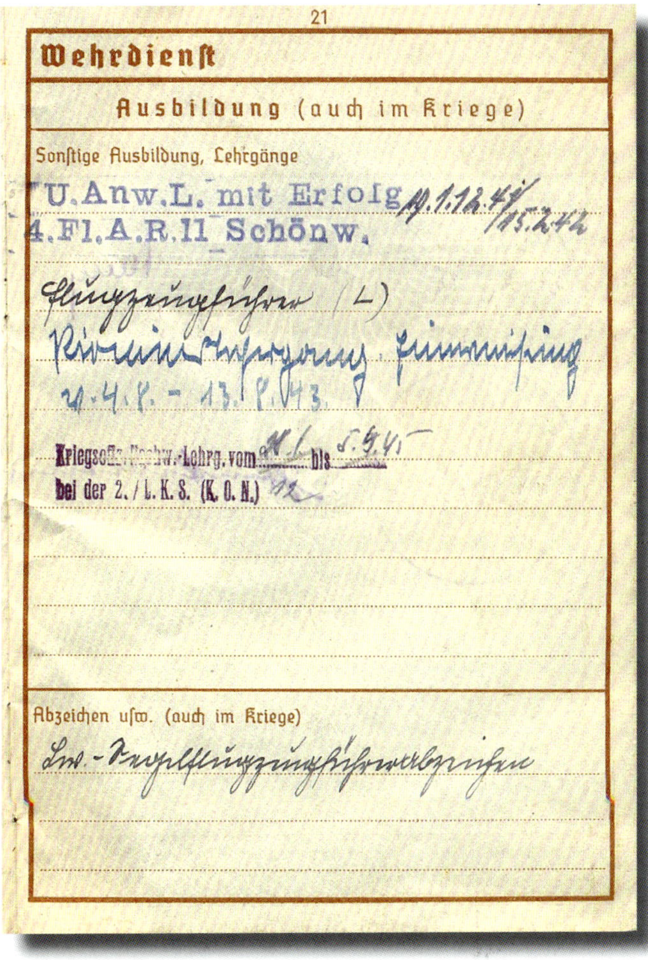

21

**Wehrdienst**

Ausbildung (auch im Kriege)

Sonstige Ausbildung, Lehrgänge

U.Anw.L. mit Erfolg 19.1.12.41/15.2.42
4.Fl.A.R.11 Schönw.

Flugzeugführer (L)

Kriegsoffz.-Nachw.-Lehrg. vom ... bis 5.4.45
bei der 2./L.K.S. (K.O.N.)

Abzeichen usw. (auch im Kriege)

Lw.-Segelflugzeugführerabzeichen

Erwin Stangl transfered in May 1944 - after some duty at Erg.Gr.(S) 2 - to an active unit, i.e., 12./LLG 1. This unit was active in the Balkan area. Stangl served as a reconnaissance flyer and was wounded on 23. September 1944. After his recovery he transferred to a war officer training course where he saw the war's end.

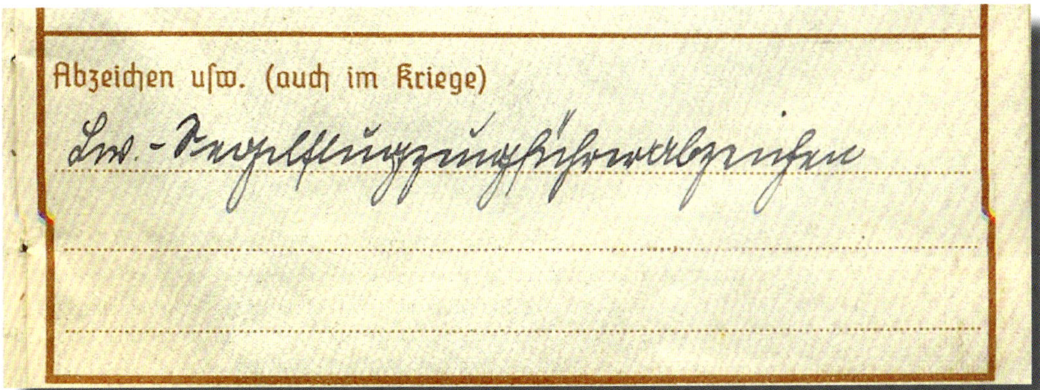

Abzeichen usw. (auch im Kriege)

Lw.-Segelflugzeugführerabzeichen

Most of the known and available service records of *Luftwaffe* glider pilots shown on page 21 exhibit the number (674) - either in handwriting or as a stamp - following the denomination *Flugzeugführer (L)* or *Segelflugzeugführer (L)*.

This number was so far only found in *Wehrpässen* of soldiers who finished the complete training and were therefore allowed to wear the Glider Pilot Badge. The presence of this number is an important indication that the bearer really was a glider pilot. In addition it is important to realize that not all service records show complete entries (as an example: posthumous awards). Known entries are:

- *L.S. Führer*
- *Lastensegelflugzeugführer*
- *Flugzeugführer (216/674)*
- *Flugzeugführer (L)*
- *Flugzeugführer (L) (674)*
- *Segelflugzeugführer (L)*

But not only is the actual title after the successful completion of the training course sometimes different, the same applies to the name of the actual badge. Known entries are:

- *Flugzeugführer-Abzeichen L*
- *Segelflugzeugführerabzeichen*
- *Segelflugzeugführerabzeichen (L)*
- *Segelflugzeugführer-Abzeichen*

There really exists a colorful mixture between theses denominations and one is well advised to check the record of the units and the training in combination with the recorded badges. This combination will reveal if it really is a Pilot Glider Badge or not.

A DFS 230 glider in rigid tow by a He 111 (France 1943).

An excellent view of the inside of the cockpit of the DFS 230 glider.

## The *Soldbuch* (Pay Book)

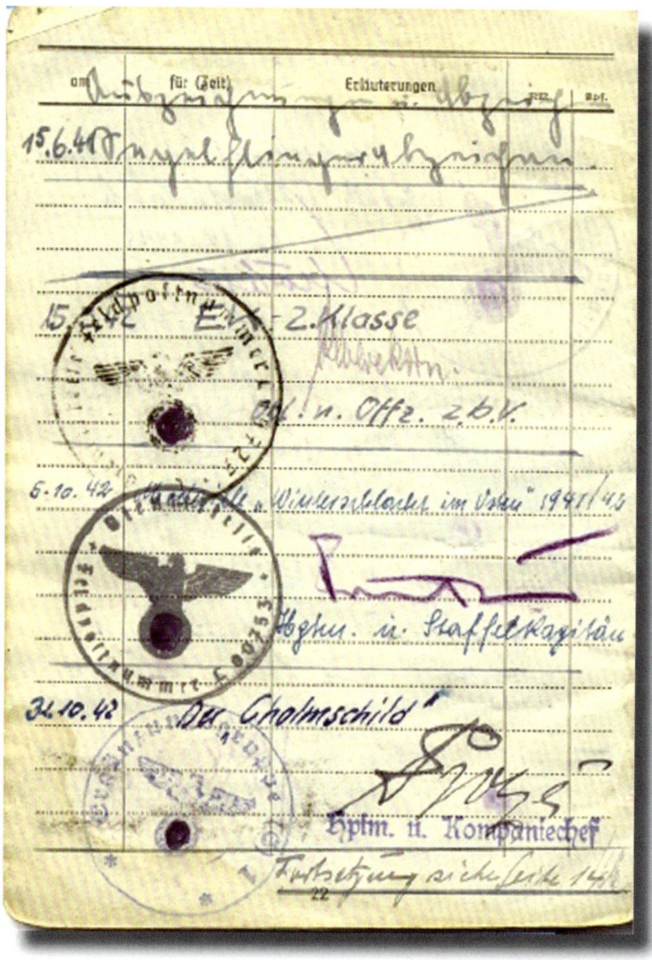
am | für (Zeit) | Erläuterungen | Rpf.

15.6.41

2. Klasse

6.10.42

"Winterschlacht im Osten" 1941/42

31.10.42 "Der Cholmschild"

Hptm. u. Kompaniechef

22

This is page 22 of the pay book of *Leutnant* R. Eck. The following orders and medals are listed:

- Glider Pilot Badge
- Iron Cross 2nd Class
- Winter Campaign in Russia
- Cholm Shield

Eck was deployed at the eastern front and conducted a supply glider mission in the spring of 1942 into the Cholm pocket.

In the summer of 1943 he was part of a glider mission to Sicily.

In 1944 Eck was promoted to *Leutnant* and transferred to the KG 200 were he was trained as an "engine pilot."

*Leutnant* Eck was chosen to conduct a special mission with a Mistel-Set (*Mistelgespann*) but time ran out. He was captured by the American forces and survived the war.

The entry of the glider pilot badge shows the same date as the award document: 15. June 1941.

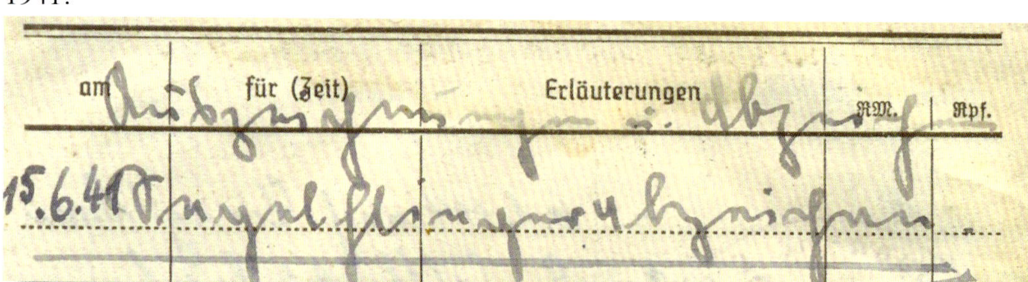
am | für (Zeit) | Erläuterungen | RM. | Rpf.

15.6.41

This picture shows page 11 of the pay book of *Leutnant* K.H. Schneider. Interestingly enough, there is no entry at all about the Glider Pilot Badge. But on this page one can find the entry:

*Spezialpersonal - Flugzeugführer (L)*

As one could observe already in the *Wehrpass* section, the wording *Flugzeugführer* is a variation of the entry *Flugzeugführerabzeichen* (glider pilot badge).

Compare the entry of the same person (K.H. Schneider) in the *Wehrpass* section page 68.

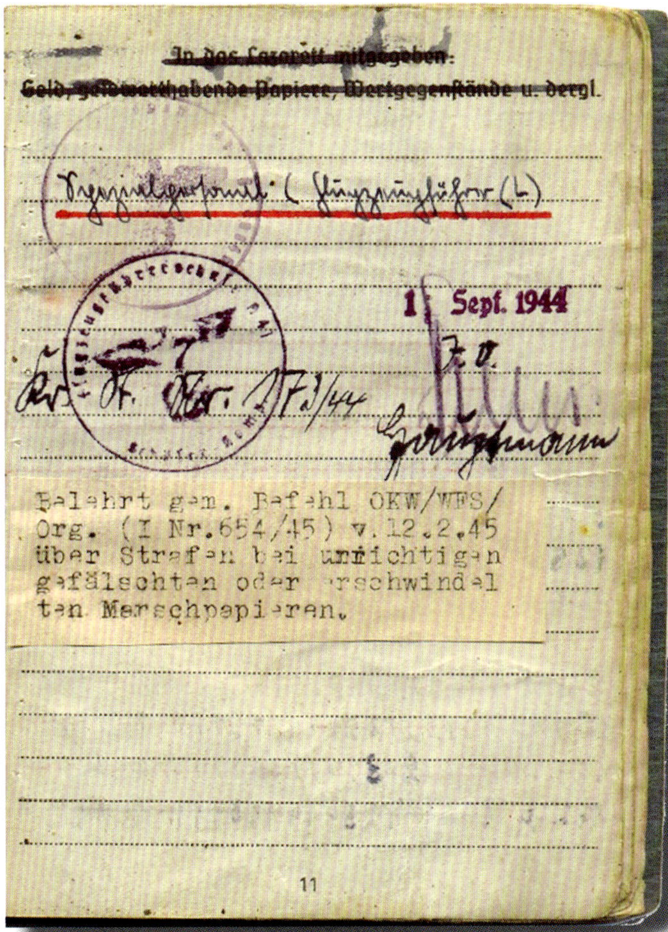


The entry *Spezialpersonal - Flugzeugführer (L)*.

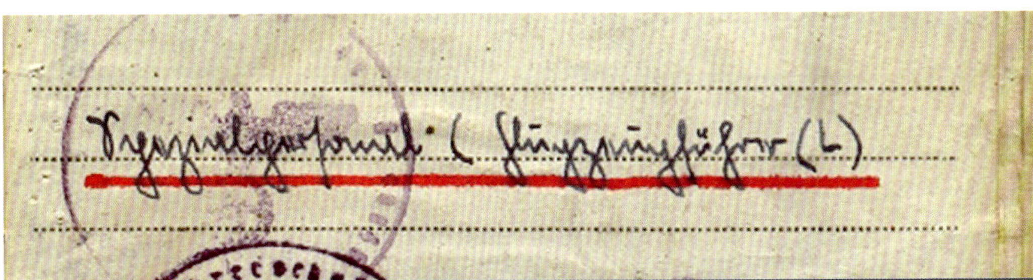

The pay book page 22 of another glider pilot. He earned the following awards:

- Glider Pilot Badge
- Iron Cross 2nd Class
- Cholm Shield
- LW Ground Combat Badge

The award entry for the Glider Pilot Badge has no date, which is unusual. But it is known that this soldier got his badge in September 1941.

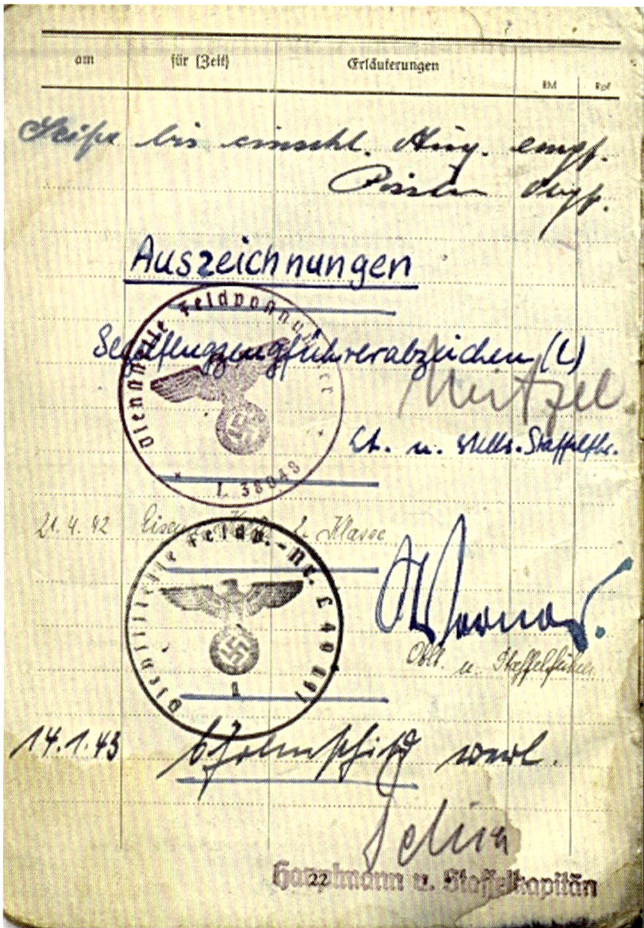


Entry for the Glider Pilot Badge, signed by *Leutnant* Meitzel, deputy Squadron Commander of *I. Gruppe Kampfgeschwader Z.b.V.1, Feldpost-Nummer L 38048*.

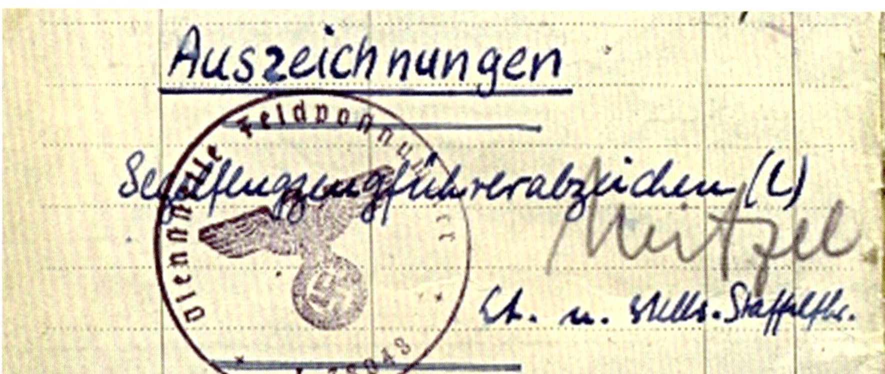

The entry for the Glider Pilot Badge in the pay book of *Unteroffizier* Hans Tost.

*Uffz.* Hans Tost was transferred in September 1941 to *Erg.-Gr. (S) 1* and was trained as a glider pilot.

In July 1942 he was moved to the *Erg.Gr.(S) 2* in Posen. This unit was relocated (without any planes) to North Africa and renamed *I. Battalion/LW Regiment Barenthin*. Between November 1942 and 13. May 1943 the unit took part in the fighting in Tunesia. During this combat Hans Tost was wounded and flown out to Italy.

For his actions in Africa Hans Tost got the following awards:

- Ground Combat Badge of the *Luftwaffe*
- Iron Cross 2nd Class
- Cuff Title "Afrika"

Not until 25. November 1943 was Tost awarded his Glider Pilot Badge. The reason was that he was not fully trained when he departed to Africa. After his recovery he finalized the requirements for the badge.

*Unteroffizier* Tost was transferred to the KG 200 after the *Ergänzungstruppen* (S) 1 and (S) 2 were disbanded in September 1944 and trained as a motor pilot. In 1945 he relocated to the tow group 1. This group, which was the last glider unit in existence at the wars end, took part in the missions to supply the Fortress Breslau.

# THE GLIDER PILOT BADGE

All of the *Luftwaffe* Glider Pilot Badges presented in this chapter are original examples of the badge which in most cases came directly from the awardee. Collectors must know that there are a lot more fakes of this badge on the market than real pieces.

Due to the rareness of the badge one can assume that there were only a very few manufacturers. As of today only two could be verified without any doubt.

The company of C.E. Juncker in Berlin - very well known for its high quality workmanship - was significantly involved in the design of the *Luftwaffe* Glider Pilot Badge, and therefore got some contracts to produce the badge. The 1941 awarded badges of Ludwig Egner and Anton Röhrle (among others) were very nicely made out of *"Buntmetall"* (literally "colored metal," in most cases either *Tombak,* which is a high copper content brass, or bronze, brass or *Neusilber*), and also showed the manufacturer stamp of C.E. Juncker.

All of the known and original examples made by C.E. Juncker have the same manufacturing features, and are therefore made from the same die. Due to the rarity of this badge it is safe to assume that different letters for a company stamp and differences in the overall configuration are signs of a fake.

Two variants exist for the badges made of *Buntmetall.* One has the manufacturer stamp. The other variant does not.There are no other differences between the two types. There also exist original C.E. Juncker Glider Pilot Badges made of zinc, and without a manufacturer stamp. Next to C.E. Juncker, *Brüder* Schneider, Wien (BSW) is the other known and verified manufacturer of this badge.

The collector has to differentiate between awarded pieces and legitimate privately bought replacement pieces.

Awarded pieces: made by C.E. Juncker, Berlin and BSW, Vienna

Replacement pieces: made by M.F.P., Posen and one unknown manufacturer

No weights and dimensions are given in this section of the book since it is the opinion of the author that such data will confuse the collector more than they are of any use. With each and every original there are minor tolerances due to the manual process. The details of the badges are the tell tale signs of an original or a fake.

## Badges made by C.E. Juncker

C.E. Juncker, located at Alte Jakobstrasse 13 in Berlin, was founded in 1851 and was mentioned in 1913 and 1916 as a manufacturer of orders and medals of all countries. In late 1944 the company was bombed and was restarted after the war for a short time.

There are four variants of the Glider Pilot Badge from C.E.Juncker, Berlin:

| | |
|---|---|
| 1. Made of *Buntmetall* | a. with manufacturers mark<br>b. without manufacturers mark |
| 2. Made of Zinc | with wreath and with detailed swastika |
| 3. Made of Zinc | with wreath and solid swastika |

The types 2 and 3 are always encountered without any manufacturers mark!

## C.E. Juncker, *Buntmetall*, Marked

This type is always well manufactured. The swastika shows signs of hand finishing and therefore differences between several examples of this manufacturer's series are to be expected.

Obverse of a Juncker *Buntmetall* badge. This badge was awarded in June 1941 to *Leutnant* Anton Röhrle.

The detail picture of the Juncker maker mark shows clearly a double stamping. The line BERLIN SW can be seen as a faint stamp under the strong one. This is not unusual with original badges of his company.

Reverse of the *Buntmetall* version of the Glider Pilot Badge from Juncker. The catch is made of round stock. The pin is round and straight and attached to a barrel hinge. The eagle is fixed to the wreath by means of two rivets.

## C.E. Juncker, *Buntmetall*, Unmarked

This type is also very well manufactured. The swastika shows signs of hand finishing and therefore differences between several examples of this manufacturer's series are to be expected. This type is absolutely identical to the marked example.

The reverse of the same badge. The construction, i.e., the hinge, the pin and the catch are absolutely identical to the marked example.

## C.E. Juncker, Zinc, Unmarked

Despite the fact that this badge is unmarked it clearly can be attributed to the company C.E. Juncker. Details of the wreath and the eagle are identical to the ones found on a marked Juncker example. The swastika is detailed.

This beautiful zinc example is attributed to *Feldwebel* Siegfried Baier.

The reverse of the same badge shows that the catch is different. It is made of flat stock which is attached to a plate, which in turn is soldered to the wreath.

The barrel hinge and the pin seem to be of the same construction as the *Buntmetall* example.

# C.E. Juncker, Zinc, Unmarked

This badge - also from C.E. Juncker - is made from the same die. The only difference from the previous model is that the area between the swastika arms is no longer cleared out manually.

The eagle has a grey patination.

The reverse is also slightly different. The catch now has a round attachment plate. This is only possible with this "uncut-swastika" Type. The omission of this finishing step could be for economic reasons.

Typically, the wash on the reverse shows the bubbles which normally appear with this type of badge.

The last picture of the Go 242 crew Köhler/Stock taken during a break before they flew the resupply mission towards Bucharest, end of 1944.

# Badges made by Brüder Schneider, Wien

The company of Brüder Schneider in Vienna (also known as BSW), a metal working shop and army supplier, was already active during the Austria-Hungary monarchy. The firm was known for several nicely executed medals of exceptional quality. During the Second World War they produced several qualification badges of the *Luftwaffe,* and held the *Präsidialkanzlei*-Number "106".

*Österreich* 8/4 1896. **Metallwarenfabrik Brüder Schneider A.-G., Wien VI./56, Bürgerspitalgasse 6—10.** 25/9 1943.

Geschäftsbetrieb: Metallwarenfabrik. Waren: Metallwaren, nämlich Knöpfe, Abzeichen, Koppelschlösser, Lederbeschläge, Orden, Schilder oder unechte Schmucksachen, Kochgeschirre, Kannen, Feldflaschen, Tafelgeräte, Obstschüsseln und Tafelaufsätze, Zucker- und Brotkörbe, Service für Mokka, Kaffee, Tee, Likör, Eß- und Vorlagebestecke, Karaffen, Untersätze für Wein-, Bier-, Wasser- und Likörbecher, Blumenvasen, Jardinieren, Keksdosen, Champagnerkübel, Bratenplatten, Bonbonnieren. (GK. 9f, 9b, 17, 19, 23, 28, 29, 30)

[43. 1795]

The picture above shows an advertisement from BSW. It is interesting to note that the scope of manufacturing went from buttons, medals, and belt buckles to pots and other cooking device, and even to cookie boxes and champagne buckets!

There exist two variants of the Glider Pilot Badge from Brüder Schneider, Wien:

1. Made of *Buntmetall* — a. with manufacturers mark
   b. without manufacturers mark

## Brüder Schneider, Wien (BSW), marked

This badge from the company of Brüder Schneider, Vienna (BSW) is completely made from *Buntmetall.* The eagle is slightly blackened.

Obverse of the Glider Pilot Badge made by Brüder Schneider, Vienna.

The crack next to the rivet (picture below) seems to be typical for this manufacturer.

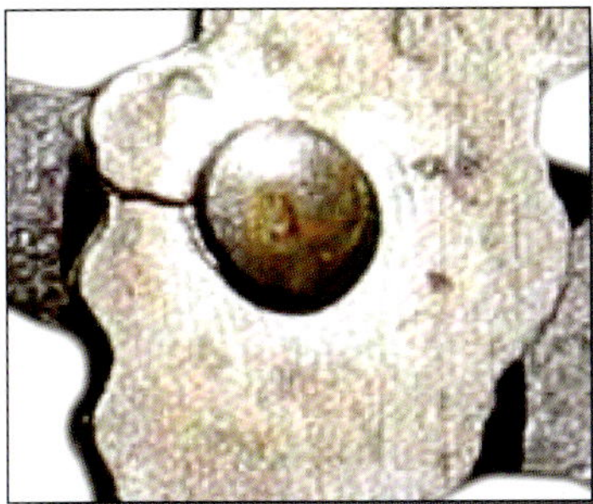

The manufacturer's mark in detail shows the initials of the company inside a cloverleaf. (right)

The reverse of the badge from BSW with the marking.

## Brüder Schneider, Wien (BSW), unmarked

This badge type is made from the same die and is therefore absolutely identical to the marked version.

This badge was awarded to *Unteroffizier* Erhard Kirsche in 1941.

The reverse shows the same cracks at the same spots as the previous badge. It appears that this seems to be a characteristic of this badge from BSW. The catch is broken due to wear.

As of this date no other than *Buntmetall* badges from this manufacturer have been observed.

The reverse of the unmarked badge from BSW.

## Replacement Badge with Larger Eagle

A few years ago Glider Pilot Badges with a bigger than normal eagle surfaced. The immediate question was, of course: "Are these badges original examples or a fantasy production?" Nimmergut asserts, in "*Deutsche Orden und Ehrenzeichen*", Volume 4, on page 2461, that such a variant did not exist during the period of the Third Reich, but that is clearly wrong. It has been proven without any doubt that such examples have been worn between 1941 and 1945.

Regarding to the pilots who had such examples, they were a privately made variant, i.e., a so called replacement badge. Not only is the eagle much bigger than the one on the official awarded pieces, but the wreath also shows unique features.

Based on the material used *(Tombak)*, one can reasonably assume that this badge variant was made in the earlier time frame of beginning to mid-1941. This contention is supported by the fact that this badge was only found by glider pilots who finished their training in the first half of 1941.

*Gefreiter* Emil Solbrig wears this variant in the picture to the left. The portrait was taken in the second half of 1942.

The detail magnification (below) clearly shows the bulky eagle.

This very rare variant also uses the usual way of attaching the eagle to the wreath by means of rivets, but there are examples known where the eagle has been soldered to the wreath, and closer investigation shows that this was not done to repair the badge but was intentionally done as part of the original manufacturing process. The reverse shows that there were no provisions at all to use rivets during the initial assembly.

Both types were verified with unshakable provenance directly from the original owners.

The obverse of the variant with the larger eagle. The badge is made of *Tombak* and in this particular badge the eagle is soldered to the wreath. This example also shows signs of manual finishing. The example is slightly worn and shows heavy patina. The badge was owned by *Feldwebel* Bruno Horn, who was awarded the Glider Pilot Badge on 5. August 1941.

It has been reported to the author by several pilots of the glider troop that this badge was acquired "privat" and was therefore never issued as an official award piece.

The reverse of this badge with the larger eagle shows the mounting of the eagle with two rivets. Please also note how thick and massive the eagle is.

The reverse also shows some remaining signs of a silver wash. The pin system is completely different to the one used with the official award badges.

A closer look at the hinge shows how well made the components are, compared to the rather crude design of the regular badge.

This unknown *Unteroffizier der Luftwaffe* clearly wears the variant with the larger eagle.

The detail magnification reveals the bulky eagle even better.

The soldier also wears a transport and Glider Clasp in Bronze which means that he conducted at least 20 combat missions. It was not easy for a glider pilot to receive this clasp, for obvious reasons.

The picture (left) shows *Uffz.* H. Behnke and was taken in mid 1943.

He wears the following awards:

- Glider Pilot Badge
- Iron Cross 2nd. Class ribbon
- Transport and Glider clasp
- large NSFK Glider Badge (one wing missing)

The Glider Pilot Badge is of the "larger eagle" variant which *Uffz.* Behnke bought as a private piece.

Magnification of the photo shows the differences to the regular badge. On can also see that the space between the swastika is not removed.

The picture (above) shows *Uffz*. Wilmskötter (right) next to his sister and brother.

He wears a Glider Pilot Badge with the larger eagle as can clearly be seen in this magnification. *Unteroffizier* Wilmskötter also has the SA Sports Badge and the embroidered version of the NSFK badge, C-Test, attached to his tunic.

This undated photo shows *Feldwebel* Alfred Kallmeyer. He has the following awards attached to his *Luftwaffe* tunic:

- Glider Pilot Badge

- C-certificate (embroidered version)

- *Hitlerjugend* badge in gold

- Romanian Pilot Badge

*Feldwebel* Kallmeyer also wears the "larger eagle" version of the Pilot Glider Badge.

Page 70 of this book shows his *Wehrpass* with the relevant entries.

The detail magnification again shows the "larger eagle" version, which can be easily recognized by the bulky eagle and the unique design of the claws.

These contemporary pictures prove without any doubt that this badge is a pre-May 1945 example, and that "non-regulation" versions of the Glider Pilot Badge were indeed worn during the Second World War.

The glider pilot's view out of the front window while under tow by a "*Tante Ju*."

## Replacement Badge by Company M.F.P.

There exists another known replacement copy of the Pilot Glider Badge. The eagle as well as the wreath of which differ to a great degree from the regular design.

This badge was privately bought by *Gefreite*r F. Stangl in the town of Posen , Prussia, right after he had finished his training in May 1943.

This variant is made of zinc. The wreath is covered with a silver wash and the eagle has a dark grey artificial patina.

The reverse of this type shows the manufacturers marking “MFP  Posen” and also shows the remaining silver wash. Notable also is the unusual pin style and the catch, which is soldered to the wreath with the help of a plate.

The magnification of the manufacturers stamp.

The question is now: Who could this manufacturer with the initial "MFP" in Posen be?

This unknown *Unteroffizier der Luftwaffe* shows the following awards:

- Glider Pilot Badge
- DRL Sports Badge
- C-Test Badge (embroidered)
- DLRG Badge

His Glider Pilot Badge shows some similarities to the badge with the "MFP" stamp on the reverse. The wreath is very similar and the eagle is also bulky.

This badge does not resemble the official badges made by C.E. Juncker or Brüder Schnei-der, Wien.

The badge also is not of the type with the "larger eagle" since the wreath is clearly different.

The Prussian town of Posen (with its airports Bednary and Schroda) was where the training group Erg.Gr. (S) 2 had its home. *Gefreiter* F. Stangl, who bought this unknown example, was trained there and also bought the badge in Posen. This allows for the following presumption:

There was a company in Posen with the name of *Metallwarenfabrik Jablonski GmbH* and one could assume that they might have used the short form like this:

**M**etallwaren **F**abrik Jablonski GmbH / **P**osen = **MFP**

But this is just an assumption, and as of today it is not proven what stands behind the "MFP".

## CLOTH VERSIONS OF THE GLIDER PILOT BADGE

Every legitimate awardee of the Glider Pilot Badge was allowed to wear an embroidered version of the badge instead of the metal version. The embroidered version could be bought from the sales department of the *Luftwaffe* in Berlin, Putkammer-Str. 16/18.

The prices for the badges were as follows:

In hand-embroidered form ......... approx. 7.50 RM
In machine-embroidered form ... approx. 0.90 RM

The difference between the two versions was in quality. The machine-woven form was for the regular troops, and the more luxurious hand-embroidered version was intended for officers.

The notion that there were no cloth versions of the Glider Pilot Badge is incorrect since the pictures on the following pages clearly show such examples in the machine-embroidered form (*maschinengestickte Ausführung*). Furthermore, both versions are mentioned in the initial regulations.

However, for unknown reason, the manufacturing of the cloth versions was forbidden as of 8. May 1942.

Original examples of the cloth version are extremely rare. The reason for the rarity should be clear: They were produced in very low numbers and were not as durable as metal versions. Hand-embroidered versions of the badge have survived, but unfortunately there is no photographic evidence in existence as of today of their wartime use. The reason is most likely that this version was not circulated very much.

In addition, there were not too many officers who were both pilots and glider pilots. Most of the officers of the *Luftlandegeschwader* 1 and 2 , of the *Verbindungskommandos* (S) and the resulting tow groups were initially "motor pilots." The very few officers who also were glider pilots, were most likely NCOs who came up through the ranks. Their version of the cloth badge was therefore machine-embroidered.

Despite the comprehensive research for this book, it was not possible to find an original example of the hand-embroidered version of the Glider Pilot Badge. Therefore, this version is not shown in this book.

## Cloth Version, machine-embroidered

The picture (left) shows the obverse of a machine-embroidered example of the Glider Pilot badge.

The example is nearly mint. The material is cotton and rayon. The badge is slightly convex due to a padding on the reverse.

The reverse of the same badge shows the padding cushion in the middle of the badge.

This example belonged to *Feldwebel* W. Schaedler.

*Feldwebel* W. Schaedler (the picture shows him still as a *Unteroffizier*) the machine-woven Glider Pilot Badge which is shown on the previous page.

In addition, he has the badge of the C-Certificate of the civil gliding proficiency badge and the ribbon to the Iron Cross 2nd Class.

The photo dates from around 7. Mai 1942.

The magnification shows nicely the effect of the padding, giving the badge a more three-dimensional appearance.

This unknown *Gefreiter der Luftwaffe* shows a cloth version of the Glider Pilot Badge together with the C-Certificate of the Civil Gliding Proficiency Badge, also in a cloth version.

This picture was taken in 1941.

The detail picture shows the color differences between the wreath and the eagle itself.

Compare also to the badge shown on page 108.

In accordance with regulations *Unteroffizier* Gustav Gronlid wears his cloth version of the Glider Pilot Badge on his *Fliegerbluse.* His metal version was awarded on 29. August 1941. The picture also shows the color difference between the eagle and the wreath, as well as the two colors used to weave the wings.

The picture to the left shows a machine-embroidered example on khaki (tropical green) backing.

Many glider troops were stationed in North Africa, Italy, and in the southern sector of the eastern front (*Krim* peninsula) where the use of tropical uniforms was normal.

Soldiers of the *Ergänzungstruppe* (S) 2 - grouped with the *Luftwaffe* regiment "Barenthin"- were fighting from mid-November 1942 until 13. May 1943 as foot soldiers in the "Bridgehead Tunesia".

Up until today, it was not known whether this version was actually issued or whether it was just a manufacturing example awaiting final approval.

The cloth badge above shows the same details and characteristics as the cloth version shown on page 108 and most likely both badges were made by the same manufacturer. Unfortunately, the name of this company is not known today.

There are several contemporary photographs from the award group of *Leutnant* Anton Roehrle which show such a badge, but only in an inconclusive way since the badge is more than half covered up.

However, another picture has surfaced (next page) which clearly shows this version in use on a tropical uniform. The name of the wearer is not known.

This unknown *Gefreiter* clearly wears a Glider Pilot Badge in the tropical cloth version.

He also displays the EK2 ribbon, the Black Wound Badge and the SA Sport Badge.

It might appear that the inner background of the badge is *Luftwaffe* blue. But the color of the private insignia on the upper left arm indicates that the background is in fact dark brown.

This is definitive proof that the tropical cloth version of the Glider Pilot Badge was in existence during the war and and was also worn.

Apart from the standard version of the cloth badge another variation shows up in contemporary pictures. This variant badge clearly shows a modified *Luftwaffe* tunic eagle which was mounted on a cloth wreath of a Luftwaffe proficiency badge.

Up to today these "hand-made" versions have only been spotted on pictures of members of the *Verbindungskommando (Segelflug) , Luftflotte 4* (VK (S) 4).

*Gefreiter* H. Schneider wears such a hand-made version of the cloth version of the Glider Pilot Badge in addition to the *Hitlerjugend* proficiency badge and the C-Certificate of the Civil Gliding Proficiency Badge.

The picture was taken sometime between November 1941 and January 1943.

The picture below of *Gefreiter* W. Gassman shows another hand-made version of the Glider Pilot Badge.

The magnification shows that the eagle is either one for the *Luftwaffe* tunic or one for the *Luftwaffe* cap. The swastika has been skillfully removed.

With the direct comparison of the two badges we have photographic evidence which shows that they are identical. This seems to be firm evidence that it is not a single unique example but that this type was worn for a short period of time by the soldiers of the unit VK (S) 4. How long that time period lasted is not known.

Former soldiers of the VK (S) 4 confirmed that the former squadron leader, *Oberleutnant* Hofmann, did not allow to wear the metal version during any mission. He therefore had the cloth versions made and distributed those amongst his unit.

# THE CASES OF THE GLIDER PILOT BADGE

Every official Glider Pilot Badge came within a special case for that award. Normally the soldier got both on the date of the award. However, for most of the awardees the case was without any value and was therefore lost over time. This is the reason that these cases are very rare today.

The case for the Glider Pilot Badge has the same dimensions as any other badge of the *Luftwaffe*. The only difference is the award description on top of the case. This could be:

**Segel=**
**Flugzeugführer=**
**Abzeichen**

or

**Segelflieger-**
**Abzeichen**

Both cases have the same dimensions, which are:

Width : 73 mm
Length: 81 mm
Height: 24 mm

(It must be made clear that these dimensions are only rough measurements and that tolerances should be expected.)
Not all contemporary cases are completely identical to each other. There are minor differences due to different manufacturers and also due to the time when they were manufactured (i.e., 1942 or 1943).

The case is made of plywood which is glued together very precisely and covered with dark blue faux leather paper. The top and bottom parts are joined together by a hinge. The front has a round-headed button. The inside of the cover is made of padded blue velvet.

The bottom part of the inside of the case holds a pressed cardboard insert which is also covered with blue velvet . Some of theses cardboard inserts are marked on the underside, but there are examples without such marking, which are also without any doubt genuine. This difference most likely lies in the different manufacturers.

The cases which have surfaced so far are to the greater part of the so-called "early model." Cases of the so-called "cheap model," i.e., the ones where the cardboard insert is covered by sprayed-on felt (as was the case with the Pilot Badge cases) , are the rarer ones. Another difference lies in the description on top of the cases:

| | |
|---|---|
| Early Case Type : | **Segel = Flugzeugführer = Abzeichen** |
| Late Case Type : | **Segelflieger - Abzeichen** |

It is not clear whether the difference in the case description is due to a difference in the time of manufacture or whether it might be due to different manufacturers.

The markings of the cardboard inserts that have been observed so far are always "626 KB / 21 /" for the Glider Pilot Badge cases. The meaning of the abbreviation is still unknown. The letters "KB" are most likely the initials of the unknown maker and the three-digit number most likely indicated the insert for a Pilot Observer Badge.

Whether the case of the Pilot Glider Badge was housed in a carton or a cellulose bag (which is sometimes wrongly called an "award bag" by some collectors), as was the case with other *Luftwaffe* awards, is not known as of today. The general usage of cheaper packing material was a result of the worsening military and economic situation of the German *Reich* from 1943 onwards. There is a very clear indication in the magazine *Deutsche Uniformen Zeitschrift* from a 1944 issue:

*"A standardized packaging for the combat badges is to be achieved. In regards to this matter the chairman (of the LDO) ascertained that the LDO always exercised the due attention to a dignified packaging for the orders and medals.*

*In recent times the packaging had to adapt to the necessities of the war economy. Therefore it will be proposed to pack in the future all combat awards in a standardized cellulose bag. The main office of the LDO will try to allocate sufficient quantities of such cellulose bags. As far as orders and medals which are still packaged in cases this can continue until the main office issues a new instruction."*

At the present time there is no indication whatsoever that any cellulose bags were used for the packaging of the Glider Pilot Badge.

# PICTURES OF RECIPIENTS

A picture showing a glider pilot with his badge or badges is always a beautiful piece of history and source of information. This chapter will show an assortment of such pictures. They all have one thing in common: they are extremely rare items.

This photo of *Unteroffizier* K. Schneider was taken in late 1941.

He wears the Pilot Glider Badge which was awarded in 1941. It is interesting to see that the badge was attached to the *Fliegerbluse* with holes through the cloth.

The usual and most practiced way of attachment was by means of loops through which the pin was pushed.

The picture was taken shortly after his participation in the Crete operation (May 1941) during which he earned the Iron Cross 2nd Class.

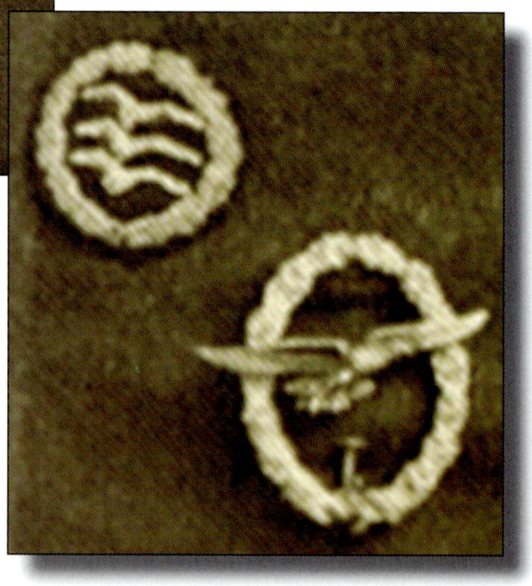

He had already earned the silver C-Certificate of the civil gliding proficiency badge on 12. August 1938.

The detail picture shows the means of attachment and also the finished swastika.

Based on the early award date one can reasonably assume that the badge was made by C.E. Juncker.

This picture shows *Gefreiter* K. Gerstmann shortly after he was awarded the Glider Pilot Badge.

He also wears the C-Certificate of the Civil Gliding Proficiency Badge.

On his lower left arm he has the insignia for flying personnel attached.

After his initial training, *Gefreiter* Gerstmann went on and earned the license to pilot the DFS 230 and the Go 242.

Gerstmann took part in the supply mission to the German *Afrika Korps* in summer and autumn of 1942 and conducted 45 front flights during this time frame. This earned him the Transport Clasp in Bronze on 1. September 1942.

He survived the war and was dismissed out of active service on 5. May 1945.

The detail picture also reveals that the badge was attached through the *Fliegerbluse.*

The finished swastika indicates another early and most likely Juncker award.

Interestingly, the C-Certificate badge sits higher than the Glider Pilot Badge.

*Feldwebel* Siegfried Baier is portrayed in this picture from the summer of 1944, shortly after his promotion to *Feldwebel* on 1. April 1944.

He wears the Glider Pilot Badge which was awarded to him on 28. August 1941. This is the same badge which is shown on page 86-87, i.e., an unmarked Juncker made of zinc.

Siegfried Baier flew several missions on the eastern front and was relocated in 1942 to the Erg.Gr. (S) 2 in Posen. There he experienced the transformation of this unit into a *Luftwaffe* infantry battalion and from November 1942 on he was deployed in the bridge head Tunis/ North Africa.

He was heavily wounded on 22. February 1943 and was awarded the Wound Badge in black.

*Feldwebel* Baier also was awarded the *Luftwaffe* Ground Combat badge, the cuff title "Afrika" and the DRL Sports Badge in bronze.

The *Luftwaffe* Ground Combat Badge was worn above the Glider Pilots Badge, which is an unmarked zinc piece from C.E. Juncker.

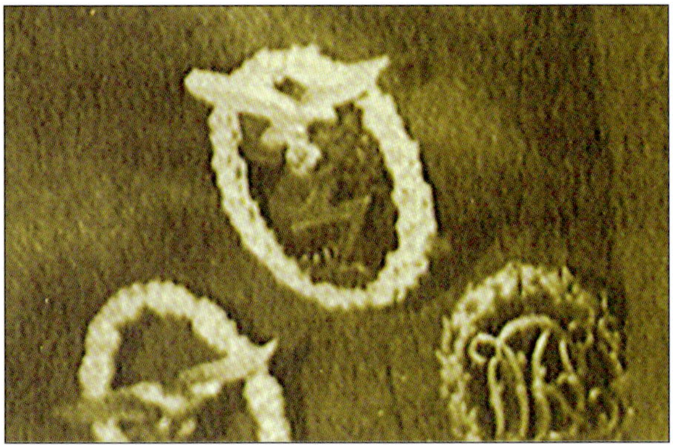

*Flieger* E. Ullrich proudly shows his Glider Pilot Badge in this picture, taken in October 1941.

The Glider Pilot Badge was awarded on 20. August 1941. He also wears the hand-embroidered version of the C-Certificate.

After the completed training *Flieger* Ullrich flew missions with a Go 242 at the eastern front.

In mid 1942 his unit was transferred to France, but in the beginning of 1943 he was back at the eastern front to fly missions to supply the Kuban bridge head, among others.

After the disbandment of most of the glider units he briefly joined a *Fallschirmjäger* unit but ended up in one of the very few existing glider units, i.e. the *Schleppgruppe 1* (tow group 1).

*Unteroffizier* Ullrich survived the war and made it back home with day and night long marches after he escaped from a POW camp.

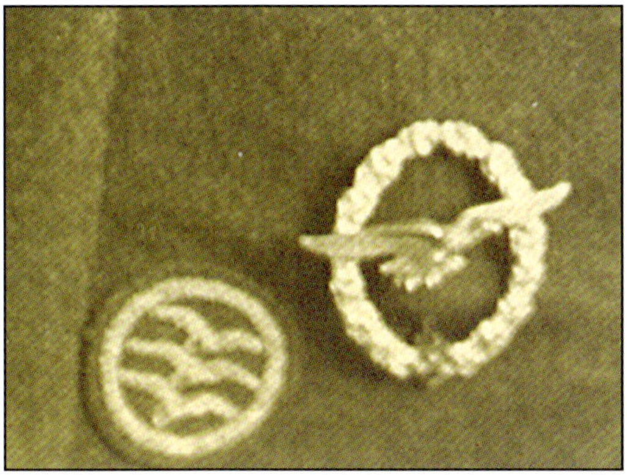

The detail picture shows the Glider Pilot Badge very clearly, but unfortunately not clearly enough to determine the maker.

*Gefreiter* Willy Jauch shortly after his training completion as a glider pilot.

Willy Jauch was transferred to the eastern front and was the first pilot to fly into the Cholm pocket on 16. February 1942.

He survived the battles around Cholm and was hospitalized afterwards for some time.

End of January 1943 Willy Jauch took part in the supply of the Kuban bridge head. During one of the missions he was forced to ditch his Go 242 in the Sea of Asow. He was saved but his flight mechanic died from hypothermia.

Willy Jauch never went back to his unit but he trained young members of the NSFK in the art of glider flight until the war ended.

The detail picture reveals that Willy Jauch's Glider Pilot Badge is an example made by C.E. Juncker.

It might be useful to compare it with the zinc model with the unfinished swastika on page 88-89.

The picture shows *Gefreiter* Gunther Oberle at the beginning of 1942.

*Gefreiter* Oberle wears the Glider Pilot Badge on the center of the breast pocket next to the hand-embroidered version of the C-Certificate.

Gunther Oberle was drafted into the *Luftwaffe* in January 1941, specifically to the II./LLG 1 in Halberstadt. After his training he transferred to the I./LLG 2 which was equipped with Go 242s.

He took part in the spring 1943 supply missions in the Kuban bridge head. He was then transferred to the dive training unit in Nancy/France and then back to the I./LLG 2 which was now stationed in Southern France. Due to an illness he could not take part in the Sicily mission

He was again transferred, this time to the *Kampfgeschwader 27,* to train the crews of HE 111 in the towing of Go 242 gliders. During the last month of the war Oberle served as a truck driver and he ended up as an American POW on 2. May 1945.

The picture (left) shows his Juncker-made Glider Pilot Badge.

*Unteroffizier* A. Eggert in an undated picture.

The author does not have enough data about A. Eggert to make an educated guess about the time frame of this picture.

Attached to his *Fliegerbluse* are the following awards:

- Glider Pilot Badge
- DRL Sports Badge
- C-Certificate in Silver (embroidered)

A. Eggert was already an active civil glider pilot before the war started and received his silver proficiency badge on 17. October 1938. It was numbered "826".

The detail picture reveals another Glider Pilot Badge made by C.E. Juncker, recognizable by the design of the wreath and the shape of the eagle.

*Obergefreiter* W. Rösler in an undated photograph, but judging by the awards the picture most likely dates from the second half of 1942.

The following awards are easily recognizable:

- Glider Pilot Badge
- Ribbon to the EK 2
- Iron Cross 1st Class
- Large NSFK Glider Pilot Badge

Rösler had award no. 1204 of the *Grosses Segelfliegerabzeichen*, instituted on 26. January 1942.

Rösler also had the Black Wound Badge, which is barely visible at the bottom of the picture.
*Obergefreiter* Rösler took part in the abortive September 1941 mission against the baltic island Ösel. Rösler never took part in any other missions. From 1942 to the end of the war he was part of the *Luftwaffen-Kommando IX* of the Horten Flying Wing development and testing facility in Göttingen.

The shape of the Glider Pilot Badge looks like a Juncker example. But as of today this has not been proven.

An unknown *Oberleutnant* of the *Luftwaffe*. The picture is interesting since the soldier wears two *Luftwaffe* proficiency badges: the Glider Pilot Badge and also the Parachutist's Badge. This is a very unusual combination, but regarding the regulations this was permissible:

*"Wearing a LW-Glider Pilot Badge together with any other Luftwaffe proficiency badge (other than the Parachutist badge) is barred."*

In addition this unknown officer also wears the ribbon to the Iron Cross 2nd. Class.

On the reverse the picture is dated "1944."

The detail picture shows nicely the Pilot Glider Badge sitting in the middle of the breast pocket. Below to the left is the Parachutist's badge.

The picture is not clear enough to determine the maker.

*Unteroffizier* H.G. Langemann wears his Glider Pilot Badge against the regulations! The picture is tough to date but it was made for sure between the last half of the year 1944 and the war's end. After the disbandment of the glider units the personnel was dispersed to several units. Some glider pilots moved on to train as motor pilots. After the successful training most of the former glider pilots were transferred to the fighter and bomber squadrons.

*Unteroffizier* Langmann shows the following medals:

- Iron Cross 2nd Class (bar)
- East Medal (bar)
- Pilot Badge
- Cholm Shield
- Black Wound Badge
- DRL Sports Badge
- Glider Pilot Badge

According to regulations, "Wearing a LW-Glider Pilot Badge together with any other Luftwaffe proficiency badge (other than the Parachutist badge) is barred."

This picture is proof that the regulations were not always followed.

An unknown *Unteroffizier* of the *Luftwaffe.*

This picture is highly interesting for the researcher of phaleristica since the Glider Pilot Badge is clearly an unknown replacement piece.

Neither the wreath nor the eagle shows similarities with the official badges made by C.E. Juncker or BSW.

This picture is proof that there are many areas that still need to be researched.

The magnified detail clearly shows that this badge was not made by either C.E. Juncker or BSW and constitutes therefore an unknown replacement maker.

Two *Luftwaffe* soldiers in front of a Go 242 glider plane.

## APPENDIX 1
## The Civil Glider Pilot Licence

As of today it is not yet known when the *Luftfahrschein für Segelflugzeugführer* was handed out the very first time. It is only certain that all of the licences are of the same appearance, They are printed in the same style. It is highly probable that private glider pilot licenses, which were issued by the several *Luftgaukommandos* (Aerial District Command), received a continuous numbering by those offices.

This license was a prerequisite to be admitted to one of the glider schools of the *Luftwaffe.*

As of today all known examples of this civil license have been found to be printed by the company Gebrüder Radezki, Berlin, SW 68.

Page number 1 of the civil license of *Feldwebel* Ulrich Sümbaum bears the number 68 and is dated 3. September 1939. The license was issued by the *Luftgaukommando VII* (Munich).

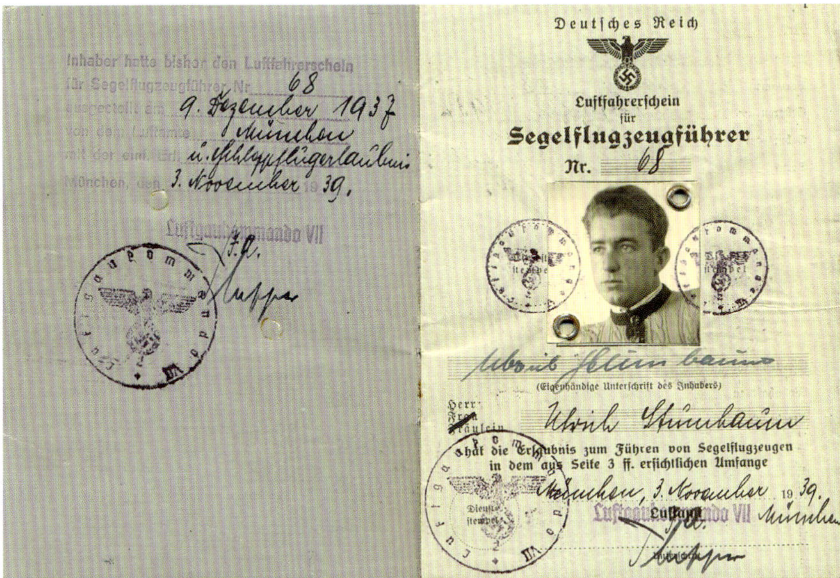
Inhaber hatte bisher den Luftfahrerschein für Segelflugzeugführer Nr. 68 ausgestellt am 9. Dezember 1937 von dem Luftamt München mit der einl. Erl. u. Schleppfliegerlaubnis

München, den 3. November 1939.

Luftgaukommando VII

I.A.

Deutsches Reich

Luftfahrerschein für Segelflugzeugführer

Nr. 68

(Eigenhändige Unterschrift des Inhabers)

Herr Ulrich Stümbaum

hat die Erlaubnis zum Führen von Segelflugzeugen in dem aus Seite 3 ff. ersichtlichen Umfange

München, 3. November 1939.

Luftgaukommando VII

Page 2 and 3 hold the private data such as the name, date of birth, place of birth and his profession.

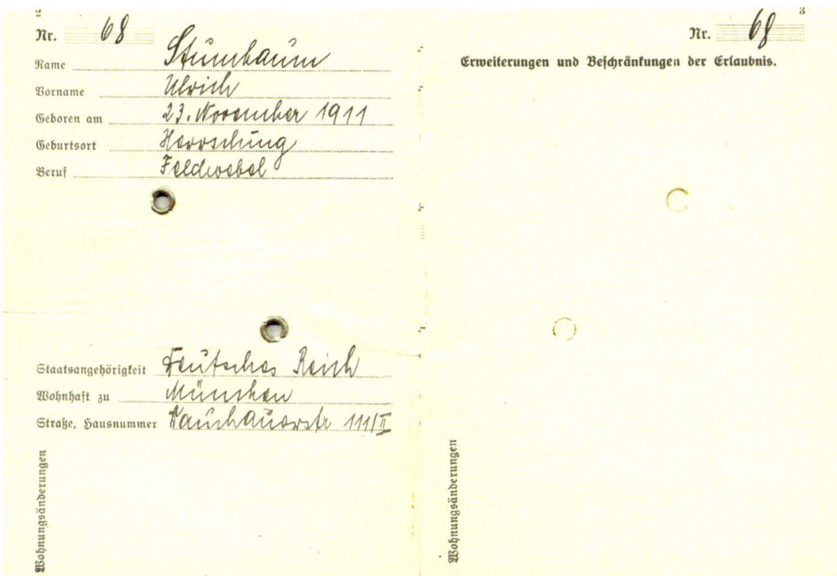
Nr. 68

Name Stümbaum

Vorname Ulrich

Geboren am 23. November 1911

Geburtsort Herrsching

Beruf Feldwebel

Staatsangehörigkeit Deutsches Reich

Wohnhaft zu München

Straße, Hausnummer [illegible] 111/II

Wohnungsänderungen

Nr. 68

Erweiterungen und Beschränkungen der Erlaubnis.

Wohnungsänderungen

# APPENDIX 2
# The *Luftwaffe* Glider Pilot License

The *Luftwaffen - Segelfliegerschein* is basically the same as the previous *Luftfahrschein für Segelflugzeugführer.* The only difference was that the *Luftwaffen-Segelfliegerschein* was issued by a military office instead of a civil one (as was the case with the previous license of *Luftfahrschein für Segelflugzeugführer)*

As of today only a few, but enough, *Luftwaffen-Segelfliegerscheine* have surfaced to recognize a certain system. It seems certain that the first licenses were issued starting 1. November 1940.

The licence was issued to the pilot by the responsible *Luftgau*. The possibility exists that a collector finds licenses which are issued by separate aerial district commands.

This license is extremely rare. As of today, only five original examples are known. The reason for that is that most recruits already had a civil license.

Page number 1 of the *Luftwaffe Segelfliegerschein* of *Gefreiter* Josef Auritsch bears the serial number 30/XVII and was issued on 9. December 1940 in Vienna.

1

Luftwaffen-Segelfliegerschein

Nr. 30/XVII

(Eigenhändige Unterschrift des Inhabers)

Name Josef Auritsch

hat die Erlaubnis zum Führen von Segelflugzeugen in dem aus Seite 3 ff. ersichtlichen Umfange

Wien, 9.12. 1940

Luftgaukommando XVII Gr. V

Hauptmann

Unterschrift

The stamp bears the inscription *Luftgaukommando XVII*, Department V. This department was responsible for the aerial supervision and civil air traffic. This also shows the close connection between the military and civil licenses.

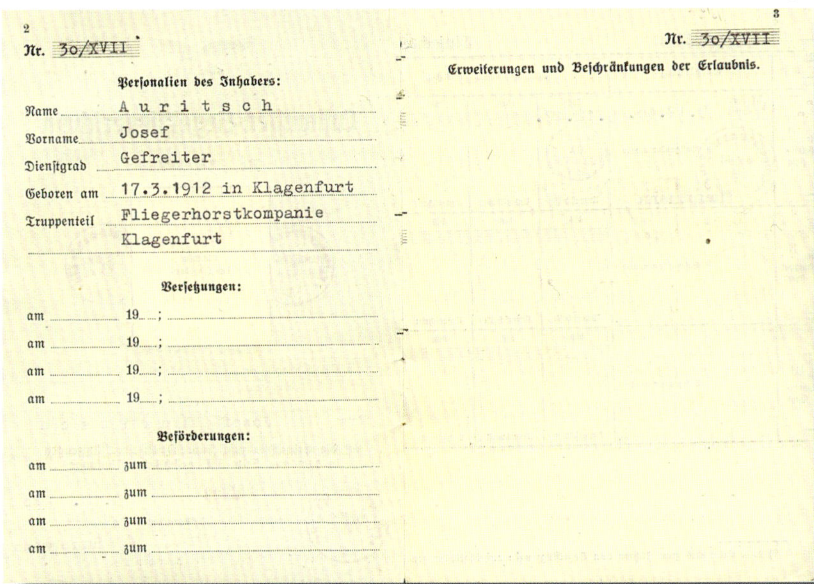
2

Nr. 30/XVII

Personalien des Inhabers:

Name Auritsch

Vorname Josef

Dienstgrad Gefreiter

Geboren am 17.3.1912 in Klagenfurt

Truppenteil Fliegerhorstkompanie Klagenfurt

Versetzungen:

am ........ 19.. ; ........

am ........ 19.. ; ........

am ........ 19.. ; ........

am ........ 19.. ; ........

Beförderungen:

am ........ zum ........

am ........ zum ........

am ........ zum ........

am ........ zum ........

3

Nr. 30/XVII

Erweiterungen und Beschränkungen der Erlaubnis.

The pages 2 and 3 hold again the personal data of the license owner.

# APPENDIX 3
## The *Luftkreise* and *Luftgaukommandos*

With the date of 12. October 1937 the territorial alignment and the numbering system of the aerial districts (*Luftgaue*) to the military districts ( Roman numerals) of the Army was performed (Regulation BLB 37, Nr. 186). The smaller aerial districts were numbered with arabic numerals.

On 1. November 1937 the following aerial districts were in existence:

| | | |
|---|---|---|
| *Luftkreiskommando* | 1 | Königsberg |
| | 2 | Berlin |
| | 3 | Dresden |
| | 4 | Münster |
| | 5 | München |
| | 6 | Kiel (for glider squadrons) |
| | 7 | Braunschweig |

There were 12 smaller aerial districts in existence in the autumn of 1937 (territorial organizations of the *Luftwaffe*) and seven aerial district command areas. On 4. February 1938 the seven *Luftkreiskommandos* were transformed into three *Luftwaffen*-groups and two *Luftwaffen*-command areas. Each *Luftwaffen*-group command area had several aerial district command areas under its authority.

Starting with the 1. August 1938 the following organisation was in place:

| | | |
|---|---|---|
| *Luftgaukommando* | I | Königsberg |
| | III | Berlin |
| | IV | Dresden |
| | V | Stuttgart |
| | VI | Münster |
| | VII | München |
| | VIII | Breslau |
| | X | Hamburg |
| | XI | Hannover |
| | XII | Wiesbaden |
| | XIII | Nürnberg |
| | XVII | Wien |

The *Luftgaukommando See* (Kiel) was disbanded on 24. April 1938 and became part of *Luftgaukommando XI* (Hannover).

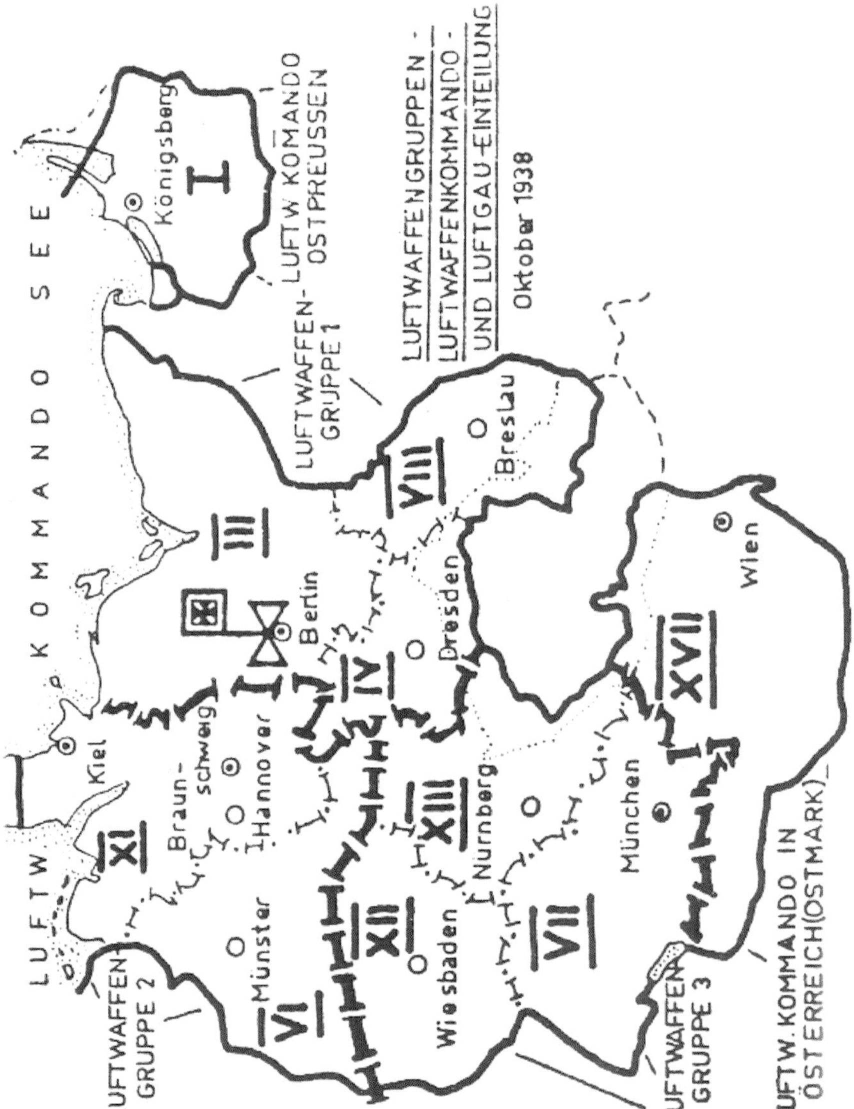


The *Luftwaffenkommandos* and the *Luftgaue* as of October 1938.

# BIBLIOGRAPHY

**Schlaug G.**, *Die Deutschen Lastenseglerverbände 1937-1945, Eine Chronik aus Berichten, Tagebüchern, Dokumenten*, Motor Buch-Verlag, Stuttgart 1985

**Klietmann K.G.**, *Auszeichnungen des Deutsches Reiches 1936 – 1945*, Motor Buch-Verlag, Stuttgart 1994

**Pawlas K.R.**, *Kampf- und Lastensegler DFS230 – DFS331*, Publizistisches Archiv, Nürnberg – ohne Jahrangabe.

**Nowarra H.J.**, *Deutsche Lastensegler an allen Fronten*, Band 42, Podzun-Pallas-Verlag, Friedberg 1978

**Pawlas K.R.**, *Die Sturm- und Lastensegler Go 242, Go 244, Go 345, P 39 und Ka 430*, Publizistisches Archiv, Nürnberg – ohne Jahrangabe.

**Binder D.**, et. Al., *Österreichs Orden, vom Mittelalter bis zur Gegenwart*, Akademische Druck- u. Verlagsanstalt, Graz 1996.

**Nimmergut J.**, *Deutsche Orden und Ehrenzeichen, Württemberg II – Deutsches Reich*, Band 4, München 2001

**Weber S.**, *Das Erdkampfabzeichen der Luftwaffe,* Verlag Weber, Schönau 2007

**Schlicht A., Angolia J.R.**, *Die Deutsche Wehrmacht – Uniformierung und Ausrüstung 1933-1945*, Motorbuch Verlag, Stuttgart 1999

*Wie es war, Die II./ LLG 1*, Halberstadt von 1940 bis 1945, Ohne Verlag, Essen 1990

Multiple discussions with Olt. Hofmann (II./L.L.G.1 and later VK(S) 4), K.H. Schneider and other members of the glider troops

*Warenzeichenblättern des Patentamtes* – B.Brücker , e-mail discussions

*Luftwaffe Verordnungsblatt* – Nr. 1635, 1940

Magazine *Schwert und Spaten*, several volumes

## Internet Resources:

www.Sammlergemeinschaft-DeutscherAuszeichnungen.com

www.Wehrmacht-Awards.com

www.GermanCombatAwards.com

www.Ritterorden.de

## Other Quality Books from B&D Publishing, LLC

### The Flak Badges of the Luftwaffe and Heer

**by**
**Marc Garlasco**

The definitive work about these two badges, including the history, documents, cases and over 50 known variants, presented in superb quality photographs.

Over 450 color pictures and over 430 pages full of information!

Also available in a Limited Leather Edition!

### The Infantry Assault Badges

**by**
**Sascha Weber**

Already highly acclaimed in its German version, this English edition is an important and indispensable addition to every collector's library.

On over 250 pages all known variants of the Infantry Assault Badges are discussed and more than 1000 pictures explain every detail of this highly collectable award. Packages and documents are also covered.
Also available in a Limited Leather Edition!

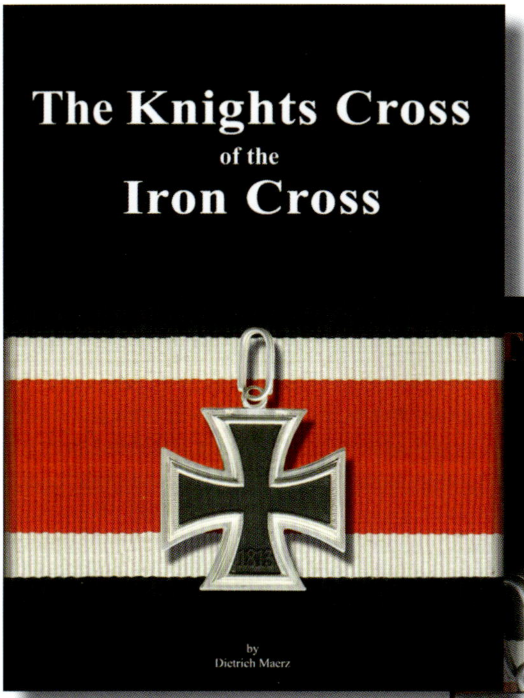


below: the new dust cover

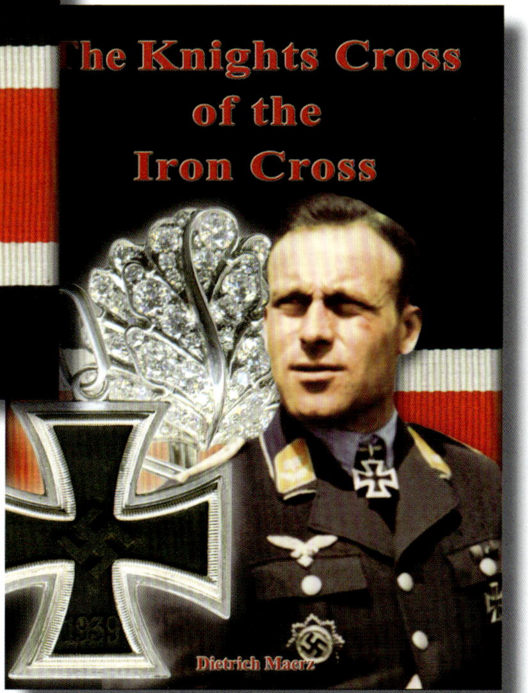


**The Knights Cross**
**of the**
**Iron Cross**
**by**
**Dietrich Maerz**

In over 440 pages the regulations, the distribution, and all accepted makers of this highest award are presented in newer before seen detail. The book also includes an in-depth discussion of the Oak Leaves, the Swords, the Diamonds, the Golden Diamonds, the Grand Cross and the Star. This book has been acclaimed all over the world and is already a classic "must-have."
Over 450 color pictures, diagrams and tables full of information! Also available in a Limited Leather Edition!

**For more information please visit: www.IronCross1939.com**
**or send an e-mail to: regular@ironcross1939.com**